T.E.B.
Doctoring in an Age of Scientific Medicine

Alan L. Graber, MD

Copyright © 2018 Alan L. Graber, MD.

Cover Photo by: Dr. Eric Dyer

All rights reserved. No part of this book may be reproduced, stored, or transmitted by any means—whether auditory, graphic, mechanical, or electronic—without written permission of the author, except in the case of brief excerpts used in critical articles and reviews. Unauthorized reproduction of any part of this work is illegal and is punishable by law.

This book is a work of non-fiction. Unless otherwise noted, the author and the publisher make no explicit guarantees as to the accuracy of the information contained in this book and in some cases, names of people and places have been altered to protect their privacy.

ISBN: 978-1-4834-8804-2 (sc)
ISBN: 978-1-4834-8803-5 (e)

Library of Congress Control Number: 2018908052

Because of the dynamic nature of the Internet, any web addresses or links contained in this book may have changed since publication and may no longer be valid. The views expressed in this work are solely those of the author and do not necessarily reflect the views of the publisher, and the publisher hereby disclaims any responsibility for them.

Any people depicted in stock imagery provided by Getty Images are models, and such images are being used for illustrative purposes only. Certain stock imagery © Getty Images.

Lulu Publishing Services rev. date: 7/30/2018

Dedication

This book would not have been written without the inspiration and support of Dotsy Brittingham. Dr.Brittingham said, "The decision to marry Dotsy was the only decision in my life the correctness of which I was absolutely certain." This book is respectfully dedicated to her.

Preface

The purpose of this book is to examine the impact of Dr. Thomas Evans Brittingham II, a legendary but sometimes controversial physician and educator at Vanderbilt University Medical Center in Nashville, Tennessee. Dr. Brittingham, known affectionately as T.E.B. by his colleagues, exerted a profound influence on a generation of young doctors during their formative years in training. He instilled in them a personal commitment to their patients that persisted throughout their lives.

T.E.B.'s relatively brief academic career, which spanned twenty-five years from the late 1950's to 1980, occurred at a time when medicine in America began to undergo radical changes. These changes were fueled by new diagnostic and therapeutic technologies, such as CT and MRI scanning, ultrasound, endoscopy, coronary artery catheterization, and chemotherapy. Simultaneously, profound changes were erupting in medical economics, increasing the involvement of both government and private business in medical practice. These changes altered the relationship of doctors to their patients; yet medicine still depends upon a doctor trying to understand and solve each patient's problems. I hope that this discussion of T.E.B. is not just a description of a past era in medicine and a physician who was an exemplar of that era, but a significant story that offers messages still pertinent to medical practice today.

I met Dr. Brittingham in 1959 when I was a third-year medical student. It was my first clinical experience, a six-week rotation on the Washington University medical service at St. Louis City Hospital. Brittingham was the newly-appointed Chief of Medicine. He taught me—by his example—how to care for sick people. He embodied in his bedside practice what it meant to be a doctor. The word *doctor* is derived from the Latin *docere*, *to teach*, but also means *to treat, to heal, to attend to*. Dr. Brittingham manifested all of these deeds. He was deeply curious about his patients' lives and felt it was an honor to serve them. His passionate thoroughness in collecting information about his patients' problems, matched with his robust intellect, worked to unravel the causes of their illnesses. Moreover, he possessed a charisma that inspired doctors in training to emulate him.

Dr. Brittingham's widow, Dotsy, told me that T.E.B. would not have wanted a book written about him. But T.E.B. was not self-effacing or unduly modest. In fact, he was genial, humorous, and captivating—sometimes theatrical. When doctors trained by T.E.B. talk about him, they exchange stories about his doctoring, how he would interact face to face with the patient. While he searched for and extracted information, he would gain the patient's confidence that this doctor understood and cared for him.

This narrative will illustrate T.E.B.'s principles and practices of doctoring. They are more than nostalgic echoes of a past age. Despite pervasive improvements in medicine, I believe that the lessons of doctoring taught by T.E.B. more than a generation ago represent the core of good patient care. They remain essential today, in an age dominated by super-specialists and never-ending advances in technology, when profit-seeking businesses view patients as markets, and when the stethoscope has become more a symbol than a tool.

Contents

Dedication v
Preface vii
Chapter 1   How to Select a Doctor 1
Chapter 2   Lambshead Ranch 6
Chapter 3   The Brittinghams of Wisconsin 19
Chapter 4   From Cape Cod to Cleveland 29
Chapter 5   Learning How To Learn at Hotchkiss 40
Chapter 6   The Most Important Decision of his Life 50
Chapter 7   White Suit Doctors 57
Chapter 8   Self-Experimentation at Barnes Hospital 63
Chapter 9   St. Louis City Hospital 72
Chapter 10   Camelot Years at Vanderbilt 79
Chapter 11   Master of Medicine at Vanderbilt 100
Chapter 12   The Doctoring Professor 118
Chapter 13   Recognition of Functional Disease 126
Chapter 14   Myths, Habits, Beliefs, and the Quirky Thunderbird 141
Chapter 15   Controversies at Vanderbilt 150
Chapter 16   Burnout at Vanderbilt 158
Chapter 17   Practice in Fort Worth 166
Chapter 18   Influence and Legacy 178

Bibliography 197
Acknowledgments 199
Notes 203

1
How to Select a Doctor

"What more delightful in literature than biography? And yet, how uncertain and treacherous is the account which any man can give of another's life."—Sir William Osler[1]

At first light on a spring day in 1972 a rusty, faded gray Thunderbird turned from Garland Avenue into the faculty parking lot. Only a few cars were there, most belonging to surgeons who must make rounds on their post-op patients before beginning a 7 a.m. case. A tall, slim man emerged and walked briskly, with a slightly forward tilt, to a side door of the hospital. He exchanged greetings with several, usually with a distinctive cordial chuckle. Everyone—faculty colleagues, interns, residents, students, nurses, cafeteria workers, housekeepers—knew Dr. Thomas Evans Brittingham II, and he addressed each of them by name. Camaraderie was a widespread attribute at Vanderbilt University Hospital.

Before the halls became cluttered with carts of breakfast trays and teams of house staff and students making morning rounds, Dr. Brittingham stopped at the bedside of several patients on the medical teaching wards, B-3100 and C-3100. He encountered interns hustling through their blood-drawing chores before breakfast. Breakfast was not to be missed, for unlimited cafeteria food and free laundering of their stiff white uniforms were the only supplements to their $50 monthly salary.

Dr. Brittingham's schedule that morning included an appointment to interview an applicant for internship. The applicant, James H. Haynes, was a thirty-seven year old chemical engineer who had abandoned that field to pursue a career in medicine. He was married and the father of four children. Haynes had driven from Duke University in North Carolina, where he would soon graduate from medical school. On arrival in Nashville, Haynes found the city jammed with thousands of tourists who had traveled from all over the world to attend Fan Fair, a weeklong country music festival. Every hotel room in Nashville was taken. He finally found a room in Gallatin, twenty miles north.

The next morning Haynes fought heavy traffic, found a place to park near Vanderbilt, and took the elevator to the third floor of the hospital. A secretary showed him to Dr. Brittingham's office. She said the professor was holding morning report with the residents but would arrive soon. Haynes knew nothing of his interviewer other than his name. He glanced around the office, trying to gain clues.

A worn, leather-covered examination table stood against one wall. Over the table hung a black-and-white framed photograph of a muscular cowboy standing next to an enormous bull. Several gray-green file cabinets were lined up next to the foot of the exam table. A window cast spring sunshine on a second table, behind the professor's desk, which held an old portable Smith Corona typewriter and a telephone. Next to the typewriter was a well-used black leather doctor's bag. A third table held a microscope and a centrifuge. The desk chair was an armed, wooden swivel type that turned easily to the phone and typewriter. A small metal desk faced the door. The desktop was bare except for an elegant wood nameplate that read "Thomas E. Brittingham, MD," and a manila folder with Haynes' name on the tab.

Two chairs faced the desk. One was a dark green leather club chair, looking soft and comfortable. The other was a light brown Windsor chair. The back and sides consisted of several thin wooden spindles attached to a solid wood saddle-shaped seat without a cushion. Its four straight legs splayed outward. Haynes knew that the Windsor chair's popularity descended in part from its use by the Founding Fathers—George Washington, Thomas Jefferson, John Adams, and Benjamin Franklin had each owned one.[2]

Haynes wondered whether his choice of chairs would be the first question of the interview.

On the wall near the door hung a small, framed letter. Haynes walked closer and read it. It was to Brittingham from the Assistant Dean of

Harvard Medical School, in 1947, informing him that his performance during the first semester of medical school was unsatisfactory and that his future in the program was in doubt. Within the same frame was his Harvard diploma; he had graduated cum laude.

Haynes contemplated the questions Brittingham might ask. Any question in medicine was fair game. Internship interviewers notoriously asked obscure questions about rare diseases to test the student's knowledge. The process was a kind of sport, and the interviewee was the prey.

Within minutes a slender man with a distinctly receding hairline breezed into the room, half-laughing, half-chuckling as he entered, offering his right hand towards Haynes with the words, "Good to see you, Jim! I'm Tom Brittingham." He appeared about six feet tall and weighed perhaps 165 pounds. His eyes, behind glasses with clear plastic frames, were gray and warm. His teeth were crooked. His fingers were long and thin. A red and blue regimental striped tie hung from his blue button-down oxford cloth shirt, with the sleeves folded above his wrists. Two chrome Cross Classic pens stuck in his shirt pocket. He wore gray wool slacks and cordovan lace-up shoes.

Brittingham seemed genuinely delighted to meet Haynes, and the younger man felt oddly comfortable. Brittingham moved toward his own chair behind the desk. He motioned for Haynes to choose whichever chair he wanted. Haynes picked the stiff Windsor chair.

"Where have you been so far?" Brittingham wanted to know. "Alabama and Emory," Haynes replied. "What did you think of Grady Hospital?" Brittingham asked. "Liked it. I liked Atlanta," Haynes answered. Brittingham looked at Haynes intently, waiting for every answer, writing down everything. Haynes was taken aback that every word seemed important to Brittingham. He had not felt such attention at Alabama or Emory, and he was beginning to feel a bit uneasy. He thought the tough question about a rare disease might come any time.

Haynes was not prepared for Brittingham's next question. "What do you think I should be looking for in an intern?" Brittingham asked. Haynes considered the question, shifted in his chair, then said, "I think you should be looking for desire and dedication."

With Haynes' answer, Brittingham leaped from his chair as if he were a cheerleader whose team had just scored. His laughter filled the room, his arms swung back and forth, he walked toward the file cabinets, then turned quickly back to his desk. He reached down and moved Haynes' file back and forth.

Brittingham sat down. He was still laughing when he said, "Well, Jim, you made a good shot in the dark. Most people tell me I should be looking for intelligence and achievement." Haynes replied, "Dr. Brittingham, everybody sitting in this chair has proved they have that already." With that answer Brittingham came out of his chair again, laughing even more. He picked Haynes' file up, put it back down, then paced around the room again, hunching slightly forward. He finally sat down again. "You're in, Haynes. You've got to get by Liddle, but you're in."[3]

After three questions, none about the science of medicine, Haynes' interview with Brittingham promptly ended. Having made his decision about Haynes, Brittingham showed him to Dr. Liddle's office and was gone.

A few minutes later, Haynes awaited Liddle's first question. Square-jawed Liddle leaned back in his chair and said calmly, "Anybody Brittingham wants gets in."

Haynes evidently *got by Liddle,* for the next week he received an acceptance letter for an internship at Vanderbilt.[4]

Almost everyone's experience with Brittingham contained, as Haynes had discovered, elements of unpredictability and surprise. As one of his students said later, an interview with Brittingham was like going on a date with a girl with a bad reputation—you never quite knew what was going to happen next. The same student called Brittingham a "slightly loose cannon," noting, however, that loose cannons can be infinitely more fun and interesting than those chained firmly to the deck.[5]

Another student said that everyone always wondered what happened to interviewees who chose the soft chair.[6]

To understand the man you must first understand his family, childhood, education, and the events in the world during his lifetime. Dr. Tom Brittingham sprang from entrepreneurial, scrappy stock: the Matthews cattle ranching empire in Texas, and the Brittingham timber enterprise in Wisconsin. Before exploring his years at Vanderbilt, we turn first to stories of his ancestors, early life, and medical training.

T.E.B. in his office at Vanderbilt

2

Lambshead Ranch

"Like most of us, he grew up to resemble and echo the people whose genes he bore"—Michael Bliss[7]

The story of Dr. Thomas Evans Brittingham II begins with a wedding on the Texas frontier. On Christmas Day, 1876, T.E.B.'s maternal grandparents, 15-year-old Sallie Ann Reynolds and 23-year-old John Alexander ("Bud") Matthews, exchanged marriage vows in the parlor of the Reynolds' home. A neighboring rancher commented, "In breeding livestock you get a certain strain. Once in a while you can see the genes mesh to make an animal of distinction. That's what happened when that little, sensitive Reynolds girl married that strong Matthews man."[8]

A guest traveling to the wedding from the east saddled his horses in Fort Worth and headed due west for 140 miles. A day's ride west of the Brazos River, he would ascend a high bluff with a view of the Clear Fork of the Brazos. Twenty-five yards wide in winter, the stream flowed rapidly over white limestone and gravel. Giant trees—pecans, black walnuts, live-oaks, and other hardwoods—lined the steep banks and bottomlands. Beyond the river lay a vast, level valley, lushly carpeted with tall green grasses, the kind that remained green all winter, and the kind that attracted Comanches to graze their horses in cold weather. Dark hills of cedar with limestone outcroppings surrounded the valley. The visitor saw the open range and herds of cattle, but no fences. The sight justified why President Franklin Roosevelt made *Home on the Range*, his favorite song, the unofficial anthem of the American West.

A wedding guest could distinguish the Clear Fork by tasting the water—the only one of the three local western tributaries of the Brazos that ran clear. Salt permeated the other two forks in the area. After descending the bluff he could traverse the river at a shallow, rock bottom crossing that Indians had used for centuries. Near the west bank he might see stone remnants of a relay station of the Butterfield Overland Mail, a biweekly stagecoach from St. Louis to San Francisco that had operated between 1858 and 1861. Any cowboy on the range that snowy day could direct a visitor to the stone house on the U-shaped Reynolds Bend of the Clear Fork, a home that Sallie's father had just built that year.

Regarding a wedding dress, shops and dressmakers were unavailable. The pioneers made their own clothes. The oldest Reynolds brother George and his wife Bettie had visited the Centennial Exposition in Philadelphia that summer and had sent white cloth—half silk and half alpaca wool—for the dress. Sallie and her sister Susan, an expert seamstress, followed a pattern from Harper's Bazaar. Susan carried out the important jobs of cutting and fitting. The wedding dress had a medium train and a wide pleated flounce around the bottom. It puffed out in the back with a bouffant effect.[9]

Except for household furnishings given to the couple by the bride's parents, Barber Watkins Reynolds and Anne Marie Reynolds, their only other wedding present was a splendid buggy from brother George. George had been a rider on the Pony Express when he was fifteen, eluding the hostile Indians who roamed at will. Years later this same George, along with his brothers Will and Phin, established the Reynolds Cattle Company, which operated in Texas, Kansas, Colorado, North Dakota, Montana, California, Utah, Nevada, and Arizona. The day after the wedding, the newlyweds drove the new buggy to the home of the groom's parents, Joseph Beck Matthews and Caroline Spears Matthews. There, friends from the surrounding country and the officers and their wives from nearby Fort Griffin gathered to celebrate the marriage.

The bridegroom's father, Joseph Beck Matthews, had come to the North Central Texas frontier from Alabama in 1858. Sallie's father, Barber Watkins Reynolds, arrived the next year. The Matthews family had known members of the Reynolds family in Alabama. In Texas they became neighbors and close friends. They were so close that two of Sallie Reynolds' older brothers, George and William, married sisters of Bud Matthews; and two other brothers, Benjamin and Phineas, married Bud's cousins. Later Sallie remarked, "It was said of us that those who were not kin to us were kin to our kinfolk."

Sallie's father, Barber Watkins Reynolds, was of English and Welsh descent. His wife, Anne Marie Campbell, was thoroughly Scottish and described herself as "rocked in the iron cradle of Presbyterianism." Barber and Anne Marie married in Alabama in 1841. Six years later they decided to try their fortunes in Texas. Barber went first, settled on an East Texas cotton farm, then sent for his family.

Anne Marie, accompanied by her two little sons, George and William, aged three and one, and a little slave girl as a nurse, boarded a boat at Wetumpka, Alabama, on the Coosa River. They continued into the Alabama River down to Mobile Bay, where they boarded a boat to New Orleans. They transferred to another boat, up the Mississippi River to the mouth of the Red River, and then west to Shreveport, Louisiana. There, Anne Marie hired a wagon team and took her seat beside the driver. Texas was a haven for lawbreakers and renegades from other states. They spent nights along the way at farmhouses. After a three week trip, they safely reached East Texas, where the reunited family settled for twelve years. In 1859 they decided to move further west. A neighbor said that West Texas was "a fine country for men and dogs, but hell for women and horses." Anne Marie Reynolds, Sallie's mother, described the trip and her first years on the frontier:

> After traveling over a week with slow teams, we landed out of sight of timber, and the first night's camp on those prairies was horrible to me, as I had been all my life accustomed, when I awoke in the morning, to hearing the sounds of axes, and the merry songs of the colored race, but lo, there was no sound to greet my ears but the howling of wolves ... while we were hunting grass, we were altogether ignorant of finding a country where the Indians depredated all the time, as they did here ... Well, the Indians raided on us for a dozen years or more, taking our stock and killing people ... They seemed to be no respecters of persons ... Our children were brought up outside of civilization, as it were. Post office

a hundred miles away, a few letters brought to us once or twice a year by chance ... no preaching ... no Sunday School ... the war commenced, and we were shut off from all supplies, but the country was covered with all manner of game, and the streams abounded with fish. If we didn't have luxuries, we always had something nice to satisfy hunger. Our boys were all good marksmen, and they killed deer and dressed the skins, and I cut and made coats and pants out of them, and the boys wore them ... they lasted well ... Our children were early in life well versed in the hardships and privations of pioneer life, and learned to depend on their own broad views and patient industry.[10]

When the Reynolds and Matthews families arrived at the Clear Fork in 1859, turkeys, prairie chickens, antelope, deer, and buffalo were plentiful. Despite dread of Indians, the pioneers had their share of festivities: feasting and dancing at weddings, square dances, and quilting parties. At one dance at a house on the bank of the Clear Fork, young men had ridden from neighboring ranches and had tied their horses near the house. When they went to retrieve the horses, they found that they had been stolen. The next morning they discovered that another dance had occurred on the river bank. Feet wearing moccasins had patted the soft earth. Before stealing the horses, the Comanches had also danced to the music of the fiddlers.[11]

In May, 1861, less than a month after the bombardment of Fort Sumter by Confederate forces, Sallie Ann Reynolds, T.E.B.'s grandmother, was born in a little ranch house. Neighboring women assisted at her birth. There was no christening, for the nearest preacher was more than a hundred miles away. Her mother, Anne Marie, then age 45, thought, "A little girl baby coming into this wild, uncivilized country ... was about the greatest calamity that could have happened."[12]

When the Civil War began, westward expansion halted, and Indian attacks increased. Many settlers retreated eastward, but the Matthews and Reynolds families stayed. They joined a few other families and "forted up" for mutual protection at Fort Davis.

After the war, the Reynolds family moved to an open valley near the Clear Fork, which Sallie called "the outside border of civilization." They lived in an abandoned rock house called the Old Stone Ranch, so named because the house and even the large corrals were built of thick stone walls. The nearest neighbor was five miles to the east. Fifteen people, including the hired men, lived in the household. Sallie Reynolds described her mother, Anne Marie, as "clinging tenaciously to the refinements of life" by covering the long table where everyone gathered for the evening meal with a white cloth. The country to the northwest was occupied for hundreds of miles by Indians and wild animals, and great herds of buffalo roamed the plains. From October to May buffalo were seldom out of sight of the house. On some days they extended as far as the eye could see. So many buffaloes were being killed that only the choicest cuts of meat were saved, usually the tongue and the hump, a tender strip between the shoulders. Hundreds of cured buffalo tongues hung in the smokehouse.

In the winter, thousands of buffaloes crowded so closely about the ranch that Sallie's mother, accompanied by one of her sons with a gun, would cut the long forelock from the buffalo heads with her scissors. She cut enough forelocks, which the family called mops, to make a nice soft mattress. Sallie inherited the mattress, and made it into pillows for cradle beds. Her two younger daughters, one of whom was T.E.B.'s mother, Lucile, inherited the tiny beds.

The following summer, Sallie's father made his yearly trip for supplies to Weatherford, Texas, on the east side of the Brazos River, leaving only two cowboys, along with his 12-year-old son Glenn and his 9-year-old son Phin, as guards. One day, fourteen Indians raided the ranch and stole five hundred head of cattle, about a hundred calves, and all the horses they could find. As it was not customary for Comanches to steal cattle, some speculated that the raiders were white desperadoes, disguised in Indian dress, complete with war paint and feathers, and with simulated war whoops. But the family who saw them were certain they were Indians. When someone asked Sallie's mother if she had been frightened, she said, "No, but I was all-fired mad."[13]

In 1867 a party of men, including Sallie's two oldest brothers, George and William, pursued Indian marauders who had stolen horses from settlers. They overtook the Indians at the Double Mountain Fork of the

Brazos, some forty miles east of the ranch, and a battle ensued. George was shot with an arrow that entered just above his navel. Arrows had great force when shot from strong Indian bows, but this arrow hit the edge of his large metal belt buckle, breaking its force. George pulled the shaft out, but the arrowhead remained in his abdomen. One of George's party swore to have the scalp of the Indian who had shot George, and he immediately succeeded. This Indian was wearing a war bonnet of eagle feathers and had a bridle studded with discs of hammered silver, presumably identifying him as a chief. George's party took several additional scalps, the white men claiming that Indians were terrified when their own tactics were used against them. For generations the Matthews family have treasured the trophies of this battle, which included the decorated bridle, beads, earrings, dozens of bracelets, bows, and quivers full of arrows.[14]

Two horses tied together, with an improvised cowhide stretcher between them, packed George home. The site of entry of the arrow healed without infection, and soon he was up and about. Somehow, the arrowhead gradually worked its way to his back, and a knot appeared near his spine. Fifteen years after it entered George's body, a surgeon in Kansas City removed the two inch arrowhead.

In the summer of 1867, the Sixth U.S. Cavalry established a fort for the protection of the Texas frontier, part of a line of defense that ran from Fort Sill in Indian Country (now Oklahoma) to Fort Concho in San Angelo, Texas. Built on a hill above the Clear Fork, Fort Griffin commanded a view of the surrounding country in all directions. Indians could not surprise this fortress. Many of the scattered ranch families moved nearby for protection and for schools. Sallie attended her first school in Fort Griffin. Her father built a house along a line of settlers' houses that extended over a mile above the fort, and the family moved there from the Stone Ranch.

On the flat below the fort a wild and wooly town, also called Fort Griffin, flourished, with saloons, dance halls, and shootings. Thousands of transients, including buffalo hunters, cowboys, horse thieves, and cattle rustlers patronized the two main activities: gambling and prostitution. There was little civil law in the area. The military took no part in civilian troubles, so that eventually white citizen vigilantes had to restore law and order. Vigilantes were iron-willed frontiersmen, tough men who thought when a man needed shooting, he should be shot.[15]

John Larn, a charming, handsome man from Alabama, resettled in Fort Griffin and married Bud Matthews' sister, Mary Jane. As such, he would be T.E.B.'s great-uncle. Local ranchers suspected Larn of cattle rustling and murder, and he associated with a tough ring of South Texas cattlemen whom the locals feared were taking over the Clear Fork range. One day a sheriff's posse arrested Larn and brought him to the jail in newly-settled Albany. That night a vigilante committee rode. Nobody is clear about the details, but Larn's body was found near midnight—he had been shot nine times.[16]

Ranches in West Texas were open range, and cattle could wander away. During the winter cattle driven south by the cold wind drifted almost to the border of Mexico. An outfit of cowboys gathered them, which sometimes required almost a month. In spring the ranchmen held general round-ups. These were spectacular affairs with several thousand cattle and a hundred mounted men weaving in and out of the herd, separating their brands. They butchered a steer, cooked five-gallon kettles of beans, and baked sourdough bread in Dutch ovens.

Sallie and Bud bought a small ranch bordering Tecumseh Creek, which enters the Clear Fork about eight miles north of Fort Griffin. Soon afterwards, a neighbor's twelve-year-old boy was scalped alive. He managed to walk home, but his injuries were so severe that he died a few months later. Another child, seven- year old John Ledbetter, disappeared from school without a trace. Eight or ten years later, a young man appeared in Fort Griffin. Dressed in Indian garb, he had the appearance of having grown up in the wild. The parents believed he was their lost boy, though he could recall nothing about his early life. When the children were young, they had played at branding cattle, and one day they had actually branded this boy. On examination he had the branding marks on his body, thus confirming his identity.[17]

The Army commanded Colonel Ranald Mackenzie, its most accomplished Indian fighter, to put an end to the Comanches.[18] Previous efforts to guide them into an agrarian, less savage life, had failed. In 1874, in a surprise attack at their hideout at the North Fork of the Red River, his troops destroyed the Comanche village and decisively defeated the hostile Indians, cornering many of them in a ravine. Then, Mackenzie and his men rounded up three thousand of the Comanches' horses, which they killed. The loss of their horses, even more than the battle itself, finally broke the Comanches' resistance. The next year they relocated permanently to a reservation in Indian Country, Oklahoma.[19]

The advance of the railroads changed frontier life. By 1880 the Reynolds and Matthews families drove their cattle to nearby Fort Worth to ship to market, rather than the previous long cattle drives to Kansas, Santa Fe, Colorado, or California. Bud and Sallie Matthews bought a brother-in-law's home, along with a large stock of cattle, near the Clear Fork. By then the extended and intermarried Reynolds and Matthews families had constructed ten homes, each of stone, on either side of the Clear Fork. They formed a partnership, Reynolds and Matthews. After all, they had been ranching together for years. They bought more land, and, with the introduction of barbed wire, fenced pastures. The land in the valley of the Clear Fork was ideal for running cattle, and the cattle business expanded.

Sallie's older brother George Reynolds, tired of the isolation of ranch life, moved to nearby Albany, county seat of Shackelford County. There he was instrumental in organizing the town's first bank and served as its president from 1884 to 1905. When he resigned because of age, Sallie's son, Joseph B. Matthews II, T.E.B.'s uncle, took over as president. Fort Griffin closed, as Indians no longer troubled the whites.

By 1884 most of the family had moved to Albany. Their ranches and cattle remained, and they added pastures formerly used by now-bankrupt sheep owners. Lucile, the sixth child of Sallie Reynolds Matthews and John Alexander "Bud" Matthews, was born in Albany on July 2, 1890. Her mother described her as "a tiny baby with brown eyes and golden curls, a little cherub."[20] Thirty-four years later she would give birth to Thomas E. Brittingham II.

Meanwhile, Bud Matthews was acquiring title to more land along the Clear Fork. He conducted his business from horseback. When he was only 19, he had been trail boss on a cattle drive from Colorado, through Wyoming, across Utah, then into Nevada. In 1894 Bud ran for the office of Shackelford County Judge. When Sallie aked him why he was going to run, he replied that he wanted to get rid of his nickname, "Bud." She had called him "Bud" since childhood. Though he only served two terms, at the salary of fifty dollars a month, thereafter he was known as "Judge Matthews."

Judge Matthews negotiated with Colonel Jessie Stem's widow in Ohio to lease Lambshead Valley. After her death, the Matthews family bought Stem's land from her heirs. Stem had been the first white settler in the Clear Fork country, arriving from Ohio in 1852, but killed by Indians two years later. His wife and four little daughters had returned to Ohio.[21] Another white frontiersman, Thomas Lambshead from Devon, England, had settled nearby. The creek which traversed the land and flowed into the Clear Fork adopted his name. Matthews established headquarters a few miles up Lambshead Creek from Stem's first outpost and called it Lambshead Ranch.[22]

Judge Matthews, a solidly-built man who always looked people in the eye, had little use for books or formal education and never graduated from high school. He took strong positions in political matters. He solidly backed repeal of Prohibition, while his wife favored its continuation. One Sunday morning, while Judge Matthews and Sallie occupied their pew in the Matthews Memorial Church in Albany, the minister began his sermon with a biblical text. As the sermon progressed, he began to preach against whiskey and expressed his support for Prohibition. Abruptly, Judge Matthews rose to his feet and stated firmly, "Brother Owen, I think it would be a good idea if you got back to your text." After that there was no further discussion of Prohibition in the Matthews Memorial Church.[23]

During later summers at Lambshead Ranch, Judge Matthews taught his twenty grandchildren, one of whom was T.E.B., to ride and handle horses, considering this instruction his moral responsibility. The grandchildren called the judge, "Other Papa," and their grandmother Sallie "Other Mama." The ranch was a second home to all of them. Judge Matthews assigned the grandsons projects such as digging post holes, building and repairing fences, maintaining the ranch roads, and cleaning septic tanks. The work day would begin with the clang of a loud bell, rung by the judge at daylight.

Sallie Reynolds Matthews was an extraordinary woman. She rode horses soon after she learned to walk, and she learned to swim in the waters of the Clear Fork, instructed by an Indian woman.[24] She was not large, and she appeared shy in public. Yet everyone who knew her marveled at her unyielding strength. Her interests were as boundless as the Texas sky.

Unlike her husband, Sallie was a lover of books and learning. During some winters she left the ranch so that her children could attend better schools in Albany, Fort Worth, or Austin. When the family had to stay at Lambshead, she educated the children herself. Despite her sporadic formal education, she was a life-long learner. One of her grandchildren once remarked, "It seemed to me that she knew everything there was to know about literature, history, mythology, the natural sciences, and almost anything else."[25] Sallie encouraged her children to attend college, and five of her grandsons, including T.E.B., attended Princeton.

When she wrote the book *Interwoven*, her modest intention was to provide a family history for her children and grandchildren. However, the book became more than a chronicle of her clan. From its publication in 1936, it became an important source in the history of the Texas frontier. Much of the information in this chapter was derived from her book.

Sallie's daughter Lucile, T.E.B.'s mother, returned to Lambshead Ranch every summer with her young children. Lucile awakened her children in the middle of the night to show them the constellations. She pointed out Taurus the Bull, the cattleman's favorite, just as her mother had done when she was a child.

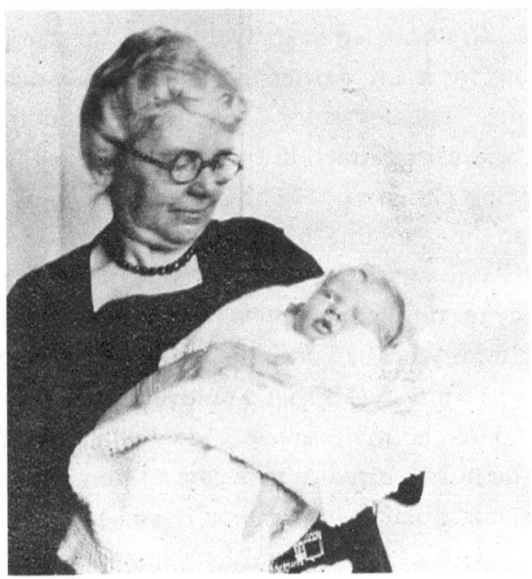

Sallie Reynolds Matthews with grandson Thomas E. Brittingham II

Watkins Reynolds Matthews, the youngest of the nine children of Sallie Reynolds Matthews and Judge J. A. "Bud" Matthews, was born in Albany and grew up on the ranch. Watt, as he was called by the family, was T.E.B.'s uncle and the last of his generation. When he was ten, he completely avoided school by working with the cowboys on the ranch, but the next year his mother imposed a strict academic schedule. He had to sit at a desk and learn reading, spelling, arithmetic, geography, history, and penmanship, while his older sisters monitored his progress. He lived his entire life on the Lambshead Ranch, except for the four years he spent at Princeton University, where he earned a degree in economics and politics. An article in the *Abilene Reporter News* quoted Watt: "Momma wanted me to go to college. She said we have plenty of cow people in this family."[26]

At Princeton both the University and the boy were wary of each other. Watt wore his hat and boots regularly. "They thought I was mighty peculiar and I thought the same of them," he said, but he adjusted. His address listed in the Princeton Class of 1921 directory never changed—"Rancher, Box 636, Albany, Texas." After graduation, except for regular trips for reunions of his class, he never saw much reason to leave the ranch.

After the death of his father in 1941, Watt ran Lambshead Ranch. Watt was famous for house parties that would draw dozens of friends for days at a time. "The best story I ever heard that defined Watt took place when he was eating lunch in the cook shack with the cowboys," fellow rancher Bob Green said. "The phone rang, and Watt talked to the individual on the other line for about 20 minutes. When he hung up, someone asked, 'Who was that, Watt?' He replied, 'Well, it was the wrong number, but they're coming for supper.'"[27]

As a young man Watt had gotten so tired of having to give up his bedroom to his parents' guests that he moved into the bunkhouse. He only moved back to the main house a few years before his death—a concession to the nurses hired to take care of him as his health failed. Until then he spent much of his time on horseback. He led the spring roundup of the ranch's 1,500 Hereford cattle and the annual group of calves, and he was the man who did the branding. The Lambshead brand is the shape of a Spanish gourd, resembling an eight which doesn't meet in the middle. Watt described the process: "I begin with a good, clean iron. I stick mine in a bucket of pork rind. The hot grease burns off dirt and hair. You want the iron grey hot, not red hot, because that'll blister a hide ... You want to make a distinct outline with that iron. The minute the hide gets pink, you stop. A lot of people burn too deep. You don't want an ugly brand on a cow she'll have to wear for the rest of her life."[28]

National oil companies drilled on Lambshead for several decades with modest success, but in the 1960's a wildcat operator from Albany struck considerable oil and gas deposits. The value of Texas ranch real estate has increased due to oil interests, far in excess of its worth based on agricultural output. Watt Matthews observed, "A pumpjack makes a fine cross with a Hereford."[29]

At Watt's funeral in 1997, 700 mourners made the 20-mile drive north from Albany, then followed the 15-mile ranch driveway to the family cemetery. Watt, who never married, was dressed in faded jeans and a Levi's jacket, a bandanna around his neck and his Stetson at his side in a plain wooden coffin. "The 5-foot-6-inch cowboy was a walking, spitting, bourbon-drinking embodiment of a bygone era," reported the New York Times in his obituary.[30] All of his siblings who survived childhood lived beyond 85, five of them until their nineties. Watt died at age 98. His sister Lucile, T.E.B.'s mother, lived to 105.

The 65-square-mile Lambshead Ranch is still owned and operated by the direct descendants of Sallie Reynolds Matthews and Judge J.A. Matthews. The cookshack is the business and social center. Its kitchen has the latest stainless steel equipment. A long steam table separates the kitchen from the eating area. In cold weather, fireplaces at each end of the cookshack burn huge mesquite logs and stumps. T.E.B. visited the ranch at least once a year. In 1974 he mentioned the proliferation of coyotes and their effect on livestock, especially calves. By then the cowboys hunted the coyotes with shotguns from the vantage of helicopters.

Photograph by Laura Wilson

Watt Matthews looking at Old Stone Ranch, where T.E.B.'s grandmother, Sallie Reynolds Matthews, lived as a child.

3

The Brittinghams of Wisconsin

"All-out effort adds a wonderful zest and sense of exhilaration to life"— T.E.B.

T.E.B. did not come from a long line of physicians. Only he and his father were members of that profession. But there was no lack of initiative, enterprise, or intelligence in the Brittingham family.

It is possible to trace his paternal ancestors back nine generations, to John Brittingham in Norfolk County, England.[31] Norfolk County, northeast of London, is a rural area of East Anglia, a region of gently rolling countryside and a coastline protruding into the North Sea. The name *Brittingham* derives from *Brettenham*, which was a village and civil parish in Norfolk County. The parish of Brettenham still exists. *Brettenham* stems from the Old English *Bretta, of the Britons*, and *ham, a homestead or village*, and means *a hamlet where Britons dwelt*.

John Brittingham's son, William, was the common ancestor of the Brittingham family in North America. William Brittingham was among the earliest settlers in Somerset County, Maryland, in 1659. William, age nineteen and single, was strong, willing to work, and eager to acquire land, which would be next to impossible to obtain in the limited social mobility of England. He wanted a *headright*, a legal grant of land to colonists. A new settler who had paid his own passage across the sea could receive one *headright* of fifty acres.

After four years, William obtained a fifty acre grant of land in Accomack County, Virginia, where he settled, married, and raised his family. His land, on Virginia's Eastern Shore of the Chesapeake Bay, bordered Somerset County in Maryland. While raising tobacco, he continued his work as a shoemaker, a trade he had learned in England. In the custom of that place and time, he owned slaves.

William Brittingham had eight children and became socially prominent. When he died in 1708 he left each of his children between 100 and 400 acres in Accomack or Somerset counties.

For the next three generations the Brittinghams were tobacco farmers in Maryland and Virginia. One served in the Worcester County Militia in the French and Indian War, and another probably served in the Maryland militia during the Revolutionary War.

The first in the family to bear the name Thomas Evans Brittingham had ambition and a good mind for business. With the opening of the American West, he picked up stakes in 1836 and headed west to Hannibal, Missouri. Hannibal was a thriving port village on the Mississippi River, 130 miles north of St. Louis. He settled there and bought two lots on Fourth Street and a tract of eighty acres in neighboring Ralls County. This tract, now a subdivision of Hannibal, is still known as Brittingham Park.

By 1844 Thomas E. Brittingham and his family were well established in Hannibal. His sons, Littleton T. Brittingham and Irvin Baird Brittingham, were the first members of the family to make careers in the healing arts. They studied pharmacy, became pharmacists, and opened a retail pharmacy on Main Street. An article in the *St. Louis Globe Democrat in 1900* described their pharmacy as the oldest retail business in Missouri.[32] Irvin, T.E.B.'s great-grandfather, remained a pharmacist in Hannibal the rest of his life.

Irvin Brittingham probably practiced some medicine as well. He kept a 64-page bound pharmacy journal, a scrapbook of remedies, in which he pasted clippings from both professional and popular magazines. The medical and pharmacy articles were pasted over original property insurance records of the Globe Insurance Company, suggesting Irvin may have been an insurance salesman as well as a pharmacist.

Irvin Brittingham's notes in his journal reveal the state of the art of medical therapeutics in the pre-scientific era of the mid-nineteenth century. He advocated treating warts with the internal use of arsenic; pneumonia with a hot poultice of chopped onions, a spider, rye meal, and vinegar applied to the chest; sore feet with a powder composed of starch, salicylic acid, and soapstone; baldness with a shampoo of bichloride of mercury, glycerine, rectified spirits, and distilled water, after bathing the scalp for ten minutes with coal tar soap; snakebite with topical application of potassium permanganate every two hours; dandruff with caustic potash; and tapeworms with eight grains of salicylic acid hourly for five hours, followed by "a good big dose of castor oil." His treatment for stammering was slowly repeating the phrase "Leander swam the Hellespont" several times a day; for cancer, ingestion of the milky soap of the Brazilian alveloz plant; for strangulated hernia, the application of hot towels to the scrotum every five minutes; for the bite of a rabid dog, vigorous washing of the site with turpentine.[33]

The notes also show an interest in science other than medicine. He kept a pamphlet from the 1880's advertising the advantages of saccharine as a sweetener; a note on the benefits of pineapple, predicting that the entire state of Florida would eventually become a gigantic pineapple grove; and a technique for igniting strips of aluminum to produce light. He even had a formula for raising a drowned body from the bottom of the water: fill a one-gallon glass jar half full of lime, then add water, stopper quickly, and drop to the bottom before the jar explodes—two or three repetitions would usually raise the body.

Brittingham Pharmacy, Hannibal, Missouri.
Left to right, Harold Brittingham (T.E.B.'s
father), Irvin Brittingham (great-grandfather),
Thomas E. Brittingham, Jr.(uncle),
Thomas E. Brittingham (grandfather)

Irvin Brittingham was well acquainted with Hannibal's most famous citizen, Samuel Langhorne Clemens. Sam was twelve years old in 1847 when his father died. For a few months he worked as a clerk in the Brittingham pharmacy to help his mother make ends meet.[34]

In 1849, Gold Rush emigrants passed through Hannibal en route to California, and eighty Hannibal residents joined them. Soon thereafter, young Clemens took a job as an apprentice typesetter in his brother Orion's print shop located above the Brittingham pharmacy.[35] He worked there for six years, learning typesetting and printing skills, before he became a river pilot. In that shop above Brittingham's drug store, Samuel Clemens wrote his first stories and humorous sketches for publication.[36] Samuel Clemens and Irvin Brittingham saw each other frequently during those days. In fact, on one of his return trips to Hannibal, Clemens, now Mark Twain, rented a room above the Brittingham pharmacy.

Irvin Brittingham married a native of Hannibal, Mary Jane League. Their first son, Thomas Evans Brittingham, who would become T.E.B.'s grandfather, was born in 1860, the year that South Carolina seceded from the Union. He grew up in Hannibal and attended private school there, then attended Hannibal College, which no longer exists. Rather than medicine or pharmacy, Thomas Evans Brittingham was interested in merchandising. The initiative and curiosity which had prompted his grandfather (same name) to leave Maryland for the West forty-five years earlier showed itself again. At the age of 20, Brittingham struck out for Clear Creek Gulch in Colorado, where he established a retail store. Subsequently he moved to California and operated a similar business.

Thomas Evans Brittingham was a smart, shrewd, honest businessman and a natural salesman. An anecdote about his first job with a lumber firm is revealing: He was anxious for a salary, but was put on commission. At the end of a month, the manager called him in and wanted to put him on a salary. His commissions had cost the company three times what his salary would.[37] His financial accomplishments would change the lives and fortunes of all subsequent Brittinghams, making it possible for his grandson to see patients in Nashville without ever charging them a dime.

In 1885 Brittingham relocated to McFarland, Wisconsin, where he opened his first lumberyard. He developed a prosperous business relationship with Joseph Morris Hixon. They established the Brittingham and Hixon Lumber Company. Brittingham served as the company's president, guiding its expansion into a chain of twenty-four lumberyards in several states. Within a few years their lumberyards, lumber mills, and timber holdings had built a financial empire and controlled a substantial fraction of the wholesale and retail lumber business in the rapidly-growing state of Wisconsin.

Brittingham moved to Madison and in 1890 married Mary Lucy Clark, a descendant of colonial New England families. She had graduated from the University of Wisconsin, a proud member of a class whose motto was "Paddle your Own Canoe." After graduation she remained active in university and alumni affairs. Thomas and Mary spent the rest of their lives in Madison, actively engaged in business, civic affairs, and philanthropy, and helped shape the city of Madison into what it is today.

Madison, the capitol city of Wisconsin, seventy-seven miles due west of Milwaukee, surrounds two lakes. The isthmus of land between Lake Mendota and Lake Monona contains the state capitol, numerous museums, major cultural and entertainment attractions, and the University of Wisconsin, which is on the southern shore of Lake Mendota.

Large lumber mills often led to the creation of company towns to serve the needs of the workers and their families. The Brittingham and Hixon Lumber Company built, owned, and operated several of these mill towns. Brittingham's attitude concerning the management of such towns was illuminated by his securing the expert pioneer librarian Frank A. Hutchins to select the books for a library for Negro workers in Alabama. The library Hutchins created was quite different from other such corporate libraries, in which businessmen would have selected the books themselves, a practice criticized for fostering a subservient mindset rather than serving the needs of its workers.

Brittingham's assets grew to 126 lumberyards, thousands of acres of standing timber, oil wells, and large interests in financial institutions throughout the United States. He was generous with his riches, especially in endowing Madison's parks. One major donation was $19,500. It established what is now known as Brittingham Park on the banks of Lake Monona, converting a stinking slough into a place of beauty. The portion of lake immediately offshore from the park is called Brittingham Bay. Between 1905 and 1908 an additional gift of $24,500 provided the park with water access, including an expansive beach, a boathouse, a waterslide, and rental swimsuits. The year it opened, 50,000 residents swam there.[38]

Though he did not attend the University of Wisconsin, Brittingham forged his own connections to the University. He was a member of its Board of Regents from 1907 until 1913 and chairman of its executive committee for three of those years.[39] In 1909 Brittingham funded the purchase of the Adolph Weinman bronze statue of Abraham Lincoln that adorns the lawn in front of Bascom Hall.

Thomas E. Brittingham and Mary Clark Brittingham,
T.E.B.'s paternal grandparents

Thomas Evans Brittingham died suddenly at sea in 1924, during a voyage from South America. Reputed as Madison's wealthiest resident at the time of his death, his will stipulated substantial endowments to both the University of Wisconsin and to the city of Madison. To mark his passing, Mayor Milo Kittleson ordered the flag at City Hall lowered to half-mast.

T.E.B.

Thomas and Mary Clark Brittingham built a Georgian-style home, *Dunmuven*, overlooking Lake Mendota and the University of Wisconsin. Young T.E.B. visited his grandparents there in summers. After Brittingham died, the family donated the mansion, fully furnished, to the University of Wisconsin. Now known as the Brittingham House, it serves as the residence of the University's president, as a meeting place for University business, and as a showcase for various art exhibits.

Dunmuven, University of Wisconsin, Madison, Wisconsin

In 1906, six generations and 247 years after William Brittingham had arrived in North America, Thomas Evans Brittingham expressed interest in the genealogy of his family. He established a family tree and corresponded with a relative in Durango, Mexico. His research is responsible for many of the details used in this chapter. Deploring the fact that no one in the family knew the name of their great-great-grandfather, he wrote that it was strange that people kept the pedigree of their horses, cows, and dogs, but did not consider the pedigree of their own family of value. He conjectured that with wealth and leisure there was time to consider knowledge of family, old family portraits, and the like. Brittingham believed that knowledge of family, if the family had no blemishes, was a source of pride and stated: "Who knows but this pride may cause some descendent to hesitate on the threshold of wrongdoing?"[40]

Scrutinizing deeds, wills, and public documents, he found evidence that his family was honest, without political ambition, and not on any prison rolls. These facts apparently provided him with satisfaction. He mentioned narrow noses and long slim fingers as common characteristics of the family.

This prosperous and generous couple were the parents of Harold Hixon Brittingham, who would become T.E.B.'s father. While Harold and his son, T.E.B., became exemplary physicians, Harold's younger brother, Thomas Evans Brittingham, Jr., T.E.B.'s Uncle Tom, pursued an equally outstanding career in business and finance and extended the wealth of the flourishing organization started by their father. In 1923 Thomas Evans Brittingham, Jr. established Lumber Industries, Inc., a financial planning and investment services business. Eventually, T.E.B. and other family members would become board members of the corporation. In 1935 Thomas Evans Brittingham, Jr. moved to Centerville, Delaware, and opened offices in Wilmington.

T.E.B.'s father, Harold Hixon Brittingham, was born in Madison and graduated from the same Hotchkiss School in Connecticut that his brother Tom, his nephew Baird, and his own son, T.E.B., would later attend. Harold, like most Hotchkiss graduates at that time, went to college at Yale, where he graduated Phi Beta Kappa. He was described by his friend Pat Glover as tall, dark and slender, with a high balding forehead, prominent nose, searching eyes, and long, slender fingers.[41] He was also known by his omnipresent sense of humor, a quality which persisted throughout his life.

Harold enrolled at Harvard Medical School, where he enjoyed great success. When he graduated first in his class in 1920, his future in medicine seemed bright. He served as an intern at the Peter Bent Brigham Hospital in Boston, under Dr. Henry Christian. The Brigham had opened in 1913, for the care of indigent persons, with a bequest from restaurateur and real estate baron Peter Bent Brigham.

Christian, the physician-in-chief at the Brigham, had been a medical student at Johns Hopkins, where Dr. William Osler heavily influenced him. Christian so revered Osler that he later edited Osler's landmark textbook, *Principles and Practices of Medicine*. Dr. Harvey Cushing was surgeon-in-chief at the Brigham. Cushing had been chief resident in surgery at Johns Hopkins and later earned the title of *the father of neurosurgery*. Cushing had already begun writing his two-volume, fourteen-hundred page biography of Sir William Osler, a book which won the 1926 Pulitzer Prize for Biography or Autobiography. When Christian and Cushing attained the opportunity to lead the residency programs at the new Peter Bent Brigham Hospital, they adopted the model of Johns Hopkins Hospital in Baltimore, which eventually became the prototype for graduate medical education throughout the United States.

"No one more fully epitomized the *personal chairman* than Henry A. Christian," wrote prize-winning author Kenneth M. Ludmerer in *Let me Heal*, an account of the residency system for training doctors in the United States. "To house officers at the Brigham, he was 'Uncle Henry,' his wife was 'Aunt Bessie,' and the house staff were 'his boys.'" Dr. Christian's former house officers kept him abreast of every detail of their lives—the birth of a child, the death of a loved one, a fire in the home, or a professional success or disappointment. One exuberant former resident wrote to him, "I can't get over the habit of telling you everything that happens to me. The latest is that I've fallen in love."[42]

During his internship year Harold Brittingham co-authored a paper on cardiac functional tests with famed cardiologist Dr. Paul Dudley White. Everything was going well for Harold at the Brigham, and he planned to continue as a medical resident there. But something happened that changed his career and his life. Harold Brittingham met and married Lucile Mathews of Texas.

At that time, there were no married house officers or residents on Dr. Christian's service at the Brigham. Teaching hospitals actively discouraged marriage among house officers, owing to the widespread belief that the work of unmarried men was more effective than that of married men. With few exceptions, an unwritten rule against the appointment of married residents persisted until the end of World War II.[43] Christian withdrew his appointment of Dr. Harold Brittingham as a medical resident, and Harold was out of a job.

ALAN L. GRABER, MD

PATERNAL ANCESTRY OF THOMAS EVANS BRITTINGHAM II

John Brittingham (1610-1654), England

↓

William Brittingham (1640-1708), immigrated
to Eastern Shore of Chesapeake Bay

↓

William Brittingham, Jr. (1682-1749), Virginia tobacco planter

↓

Poynter Brittingham (1726-1773), French and Indian War

↓

Thomas Brittingham (1750-1825), Revolutionary War

↓

Thomas Evans Brittingham (1773-1876), moved
from Maryland to Hannibal, Missouri

↓

Irvin Baird Brittingham (1825-1919), pharmacist in Hannibal

↓

Thomas Evans Brittingham (1860-1924), Wisconsin
lumber empire and philanthropist

↓

Harold Hixon Brittingham (1893-1937), Physician in Cleveland, Ohio

↓

Thomas Evans Brittingham II (1924-1986), T.E.B.

4

From Cape Cod to Cleveland

"I have such a deep feeling of how much of me has come from you." —T.E.B., in letter to his mother, February 16, 1966

In the summer of 1921, near conclusion of his internship year, Harold Brittingham and a few friends rented a beach house on Cape Cod for a short holiday. By chance, a young woman named Merle Reynolds Harding and her husband were renting the house next door. Merle, a friendly, gregarious Texan, discovered that her neighbors on the beach were bachelors and Harvard doctors. She called her unmarried cousin, Lucile Matthews, who had graduated from the University of Texas at Austin and then returned to the Lambshead Ranch, informing her what was going on. Lucile apparently decided that this would be a good time to visit her cousin and promptly departed for Massachusetts on the next train.[44]

Evidently Harold and Lucile hit it off on Cape Cod. A proposal of marriage followed within a few months. One of the first letters Lucile received, from Harold's mother, Mary Clark Brittingham, announced that Harold had told her he had found "The dearest and sweetest girl in he world"[45]

Lucile's brother Watt Matthews had been to the rodeo in Albany watching the bronc riding. He wrote to her from Lambshead Ranch that Dr. Brittingham "surely landed on his feet when he got you."[46]

The wedding took place on November 8, 1921, in Fort Worth, Texas. A partial record that Lucile kept of 103 wedding gifts listed silver bread and butter plates, a coffee service, several hand painted porcelain plates and dishes, a few checks, a Bible, and a registered calf. They brought the silver and dishes to their first home in Brookline, Massachusetts. The calf remained in Texas.

Old friends of Harold Brittingham from Wisconsin, then living in Cleveland, realized that Harold had no job. They knew that Dr. Howard Karsner at Western Reserve University was looking for an assistant in his research laboratory and suggested that Harold apply for the position—and that he also explore the possibility of a later affiliation with Dr. Roy Scott at Metropolitan General Hospital. So, early in 1922 Harold and Lucile relocated to Cleveland, where Harold worked in the laboratory. His title at Western Reserve University that year was *Demonstrator in Physiology.*

In January the following year Lucile gave birth to their first son, Harold Hixon Brittingham Jr. The infant lived only three days. He was the first Brittingham to be buried in the family plot at Forest Hill Cemetery in Madison, Wisconsin. The night before burial his casket rested in a bedroom reserved for grandchildren at *Dunmuven*. The grave was surrounded by snowy evergreens, with the ice of Lake Mendota visible in the distance.

Lucile was soon pregnant again. The baby, our T.E.B., was born February 6, 1924, and named Thomas Evans Brittingham II, in honor of his grandfather.

Tom was cherished by his parents, lessening, to some extent, the lingering grief of his older brother's death. They cuddled him, dressed him in the finest clothes, and made him the center of attention. In his later years Tom would describe his life in Cleveland as that of "a pampered poodle." He said that he would never have grown up had it not been for his war experience in the United States Army in the Philippines.

Tom's sister, Sally, was born two years later. They were always close, and many years later Sally would play a central role in what Tom called the most important decision of his life.

By this time, Harold Brittingham had joined Dr. Scott at Cleveland Metropolitan General Hospital (now MetroHealth Medical Center). There Harold devoted himself to problems at the bedside. He quickly assumed an increasing burden of clinical teaching and administrative responsibilities. Scott often said that he wanted Brittingham to succeed him at the hospital and medical school after he retired.

Britt, as he was called, also opened an office for the private practice of internal medicine, along with two old friends from Madison, Donald M. Glover and Donald Bell, and James V. Seids, all surgeons. This arrangement of an internist sharing an office with three surgeons may seem unconventional today, but it worked for them. Their association was soon interrupted, however, for in 1927 Harold developed pulmonary tuberculosis. There wasn't much doubt about how he contracted tuberculosis. He saw many patients with tuberculosis at Metropolitan General Hospital. Thoracoplasty as a treatment for tuberculosis was a special interest of the surgeons there and attracted large numbers of patients with advanced tuberculosis.[47]

Harold was treated at the Trudeau Institute at Saranac Lake, New York for two years. His wife Lucile, three-year-old Tom, and one-year-old Sally accompanied him there. They lived in one of the cure cottages, where patients with tuberculosis rested in the fresh air of the Adirondack Mountains. Elsa Foerster, the children's nursemaid, accompanied the family, risking her own health. She was fond of the children and a great influence on them.

The two years at Saranac Lake were not entirely bleak. Between long hours of rest, Harold, Tom, and Sally went sledding, built snowmen, and tried skiing. In warm weather they swam in the lake.

During their time at Saranac Lake Harold and Lucile conceived a third child, Mary, who was born shortly before they returned to Cleveland. As they were leaving, the train whistle blew unexpectedly and frightened Tom and Sally, but not the baby. Harold realized immediately that Mary was not a normal child. She was diagnosed with Down Syndrome.[48] The young Brittingham family suffered their third heartache (after previous death of infant Harold Jr. and life-threatening illness of Harold) when Mary was sent to the Devereux School in Paoli, Pennsylvania, where she lived for more than fifty years.

When Harold Brittingham returned to Cleveland in 1929, he resumed his positions as Assistant Head of the Department of Clinical Medicine at Metropolitan General Hospital and Assistant Professor of Clinical Medicine at Western Reserve University School of Medicine. Tom's younger brothers, John Matthews Brittingham and Robert Scott

Brittingham, the latter nicknamed Jigger, were born in 1930 and 1932, respectively.

Tom began first grade at a public school, Roxboro Elementary, in Cleveland Heights. He may have been disconcerted at the onset, after seeing his father's life recently threatened by tuberculosis and a new sister sent away with a disease he did not understand.

Tom Brittingham, left end of 3rd row, with his 4th grade class at Roxboro Elementary School

The Brittinghams and their children traveled to the Lambshead Ranch every Christmas. Family ties, both immediate and extended, were important to Lucile Brittingham, and she took every opportunity to expose her children to their Texas relatives and to life on the ranch.

The family spent much of each summer at Martha's Vineyard, visiting with the Elliott Cutlers. Harold and Elliott had been friends at the Brigham in Boston, where Harold studied internal medicine and Elliott eventually became the protégé of Harvey Cushing in surgery. A photograph taken about 1928 shows four-year-old Tom and Elliott Cutler in a boat sharing a cigarette. Harold visited a week or two at a time, but

Lucile kept the children at Martha's Vineyard for many weeks most of the summers of their youth.

Harold was an enthusiastic tennis player. Tom, with encouragement from his father, played a lot of tennis on Martha's Vineyard. In a letter to his father, written at age eleven, he indicates how conscientious he was about improving his game. He boasted that he could handle the American Twist, topspins, slices, and chops; that his service was fast and deep; and that he did not run around backhands. He also requested permission to go to Newport, Rhode Island, with his tennis instructor and a few friends to see Davis Cup players perform. The trip would cost less than $5.00

Tom received permission for the trip to Newport.

At the Vineyard, Tom met some of his father's medical colleagues, who visited for a few days at a time. Cutler himself was renowned for performing the first mitral valve commissurotomy.[49] Tom was exposed to some of the most eminent physicians of that time, but as a child he knew them as ordinary individuals, not by their credentials. One can speculate that his familiarity with these men in the relaxed setting of Martha's Vineyard contributed to the mistrust of academic titles he held so forcefully as an adult.

Harold Brittingham made rounds three times each week with the Case Western Reserve interns who rotated through Metropolitan General Hospital. Robert N. Buchanan Jr., later a respected senior dermatologist at Vanderbilt, was one of the thirty-six interns. Dr. Buchanan described Harold Brittingham as a knowledgeable, keen clinician, and a perfect gentleman with beautiful manners.

Harold twice invited Dr. Buchanan to his home for dinner before an evening at the Cleveland Symphony. Dr. Buchanan remembered the Brittingham house as a formal and ordered place with a butler and a cook. Lucile, the perfect hostess, presided over several well-mannered children at the table.

Many years later Dr. Buchanan considered Tom, who by then was his Vanderbilt colleague, as "The noble son of a noble sire."[50]

Tom spent the sixth through ninth grades at the Hawken School, a private school a few miles east of the Cleveland suburb of Shaker Heights. Its founder and first teacher, James Hawken, was the guiding spirit of the school. He was a disciple of the educator and philosopher William James. The school's motto, *Fair Play*, was posted in every room.

The Hawken School enforced high standards on boys both wealthy and smart enough to attend it. Each year a *Character Report* on each student was prepared by a faculty member. These reports covered not only academic performance and moral character, but also social and athletic development. Copies were sent to the student's parents and the Headmaster. Tom's first report, written by Hiram Collins Haydn II, described his wide range of intellectual interests, his enthusiasm, and his sensitiveness to others. The report then shifted to athletics, noting Tom's slight build and previous inexperience in playing football. The report stated that his mettle was severely tested in athletics, but in playing baseball he had "the stuff."[51]

The next year's report commented on Tom's proficiency in the schoolroom and noted his encyclopedic knowledge of athletics facts and figures. It concluded "There is a substantial quality about him that is comforting in a very unsubstantial world."

Tom also had annual detailed growth and developmental assessments, arranged by his father, at the Brush Foundation at Western Reserve Medical School. At age eleven he had the intelligence of an average seventeen-year-old boy, as judged by the Otis test.

All seemed well for the Brittingham family in the summer of 1936. Tom, age twelve, had finished his second year at the Hawken School and returned to Martha's Vineyard to play tennis. Sally was ten, John was five, and Robert was two years old. Harold was forty-two, engaged in a busy private medical practice, and positioned to succeed Dr. Roy Scott as Chairman of Medicine at City Hospital. He and Lucile had moved to a comfortable home in Shaker Heights.

Tom wrote to his father in July that the tennis club at Oak Bluffs had saved money for a ping-pong table by putting together two tables they already owned; it was just about regulation size. This was the last letter he ever wrote to his father.[52]

Later that summer Tom's father experienced a grand mal seizure and was found to have a brain tumor. After consultation with physicians in Cleveland, he traveled to the Montreal Neurological Institute to see Dr. Wilder Penfield, a former fellow house staff physician at the

Brigham. Penfield, trained by Dr. Harvey Cushing, was the pre-eminent neurosurgeon in North America at that time. Dr. Penfield performed surgery, but the seizures continued. After the operation, Harold went into a rapid decline. He spent the Christmas holidays in Texas, but in early January he was in a hospital in Cleveland, near death.

Tom, in the only personal diary he ever kept, recorded his daily activities, thoughts, and the weather for the month of January, 1937. The diary may have represented a coping mechanism suggested by his mother.

On January 1 twelve-year-old Tom wrote in his diary, "Dad is still alive but it is only a matter of days. Aunt Ethyl,[53] Aunt Margaret,[54] Uncle Watt[55] and Uncle Tom[56] are here. Fair-warm."

The next day Tom wrote, "Dad was conscious. Watt took me to a movie which was very funny ... Rainy-cold."

On January 6, 1937, he wrote, "Dad about to die ... Aunt Peg[57] arrived today ... Cold."

That day, Harold Brittingham died at St. Luke's Hospital in Cleveland

The next day Tom wrote in his diary. "Dad died last night at 11 p.m. He will be buried in Madison tomorrow. It sure is a damned break. Rainy-warm."

Tom's last diary entry on January 30 states, "I went to a swell party that lasted until 10:30 pm. I won a prize. I got a new girl."

By the end of 1937 a group of friends and colleagues, led by Dr. Robert M. Stecher, had organized materials and secured funding for the establishment of the Harold H. Brittingham Memorial Library at Metropolitan General Hospital. At the formal dedication of the Library over two hundred colleagues and friends, many in formal attire, attended.[58]

The library is still active.[59]

Harold's memory lingered in those who knew him long after he died. More than twenty years after his death, his friend, attorney David K. Ford, remembered him in an address delivered at one of the library's functions. Ford recalled Harold's smile, stating that when Harold entered a room, the distresses of the body and soul receded. He described Harold's tenderness, optimism, and intuitive reasoning as the "alchemy of good health." He showed how Harold reassured a patient that nothing was insoluble. And he portrayed Harold as a relentless questioner who raised the standard of thinking of those about him, concluding his address by reminding his audience that one debated with Harold at peril, because he was always fortified with facts.[60]

Dr. Harold Hixon Brittingham, T.E.B.'s father

These same qualities were later expressed vividly in his son, Dr. Thomas Evans Brittingham II. No one influenced young Tom in his childhood more than his father, Harold Brittingham.

After Harold's death, Lucile found herself alone with four children, far from her roots in Texas. She traveled frequently with them and poured herself into their education. She never remarried.

Tom's tentative thoughts of becoming a doctor were confirmed by Harold's death. Lucile said that that Tom realized he was now the man of the family and would become the one to wear Harold's shoes. He would certainly become a doctor, but he could not choose then between medicine and surgery.

Lucile wrote of Tom's decision to Dr. Donald M. Bell, a close friend of Harold and one of his practice associates. Dr. Bell was undoubtedly motivated to fill in some way the void left by Harold's death, and wrote a letter to Tom, advising him about his future education. Bell, a surgeon, advised Tom that it made no difference whether one practiced medicine or surgery, the goal of helping people was the same. Surgery, Dr. Bell wrote, was merely a highly specialized form of treatment requiring extra years of training to develop technical skill. He counseled that extracurricular reading and study was as important as that assigned in school. Age thirteen was the time to broaden one's education; the time for depth would come later. Bell reminded Tom that outdoor play and exercise was just as important as a college education. He did recommend several specific books to read, books which told the stories of great men and how they achieved their greatness. Among them were Vallery Radot's *Life of Pasteur*, Sir William Osler's *Aequanimatas and Other Essays*, and the two volume set, *Memoirs of Ulysses S. Grant*. The latter Tom could find in his grandfather's library, "in the book case in the upstairs hall about in the middle of the shelf that will strike your middle vest button." But he also recommended reading a lot of good detective stories, including Sherlock Holmes, because "a good doctor must be a pretty fair detective."[61]

Thirty years later, many Vanderbilt students and residents would attest that T.E.B. had followed the advice to read detective stories.

During the summer of 1937, Uncle Walt Matthews brought Tom down from Ohio to work as a cowboy on family ranches near Albany, Texas. Tom learned to round up cattle with the other cowboys. Once he became so thirsty out on the range that he pushed the cattle aside and drank water out of one of the stock ponds. Tom and his cousin, Watt "Palo" Casey, Aunt Ethel's son and two years his senior, worked on Lambshead Ranch and on what is now known as the Phin Reynolds Ranch. They each were paid a dollar a day.

Watt had the boys out of bed by 5:30 every morning. They herded cattle, sometimes riding their horses for twelve hours. They carried a lunch of apricots and biscuits, as their work often took them far from ranch headquarters. When cattle didn't need herding, Tom and Palo repaired fences.

Tom Brittingham treated his first patients that summer. He took an interest in treating newborn calves for screwworm. This involved spraying the calves' navels with a disinfectant to prevent the screwworm larvae from infesting them. Screwworm infestation produced a characteristic odor, and Tom became adept at locating infected calves in the herd.

At the end of a day's work Tom and Palo swam in a pond on the ranch known as *Trout Tank*. Once they drove into Abilene to bowl. They shot at prairie dogs and jackrabbits and plinked a few cans with a .22 rifle. Once they killed a big rattlesnake with a volley of rocks. During that summer T.E.B. played his first known round of golf, a sport he would not pursue. Tom and Palo borrowed a set of never-used clubs from Uncle Watt and played nine holes at the weedy public course in Albany.

Palo, who would become a veterinarian, saw his 13-year-old cousin as a boy who was fun-loving, very intelligent, entertaining, honest, quick-witted, dependable, kind, and a good companion who joked a lot. He was completely unpretentious about his family's wealth, unsentimental, at least outwardly, about his father's death, and a pretty good horseman. He read a lot at night. Palo knew that Tom enjoyed his work on the ranch, but that ranch work was not what he wanted to do in life.

T.E.B.(right) with his cousin, Watt "Palo" Casey, working as cowboys on Lambshead Ranch, 1937

By late August Tom returned to Cleveland and began his fourth year at the Hawken School. The final Character Report from Hawken mentioned one quality that needed attention—a tendency toward intellectual intolerance, which would be pertinent to his adjustment to Hotchkiss the next year. The Character Report suggested simply keeping before him the notion that such use of intelligence was not wicked, but was beneath his dignity.[62] Tom graduated from The Hawken School in spring of 1938, second in his class.

That summer Lucile took Tom, her other children, and two cousins on the train to Carpinteria, California. She rented a house on the shore for several weeks. The children, ages three to sixteen, spent much of their time playing on the rocky beach. Tom played tennis and took several tennis lessons from the professional at the Biltmore Hotel in nearby Santa Barbara. Betty Burns Densmore, one of the cousins, remembered that Tom consumed the sports pages every day, and that he was in command of all the baseball statistics.[63] He was particularly interested in the accomplishments of Bob Feller, a pitcher for the Cleveland Indians.[64]

One moonlit evening at high tide Tom led the children grunion fishing. Grunions are small fish that swarm onto California beaches at night to spawn. Tom became very excited scooping up the fish in his bare hands and led the group a hundred yards or so down the shore to the beachfront of Irene Dunn, an Academy Award-winning movie actress. Ms. Dunn approached the children and asked what they were doing on her property. Tom enthusiastically told her and proceeded to show her how to fish for the grunions. Irene Dunn soon was barefoot in the Pacific next to Tom, catching grunions with her hands.[65]

Dr. Tom Brittingham's enthusiasm as a teacher, well known years later at Vanderbilt, was already present when he was a teenager. His enthusiasm, whatever the subject, could stir up a group and get them moving in his direction.

Within a few weeks Tom was on a train, headed to the Hotchkiss School in Lakeville, Connecticut.

5

Learning How To Learn at Hotchkiss

"If you want to have fun, there's no place like a creek"—T.E.B.

The Hotchkiss School is situated on wooded farmland overlooking the Berkshire Mountains, fifty miles northwest of Hartford, Connecticut. Benjamin Berkely Hotchkiss, *Berk* to his friends, was a world-famous artillery engineer who made a fortune during the Civil War selling armaments and munitions to both the Union and the Confederate armies.

In 1850, Berk married Maria Bissell, a well-bred farmer's daughter and schoolmistress. After the Civil War, Berk and Maria's marriage unraveled. Berk wanted a son to inherit his business, and Maria was unable to conceive. He also expected his wife to entertain his business associates, but Maria, though not a recluse, was shy and not a naturally gifted hostess. Berk had already taken a mistress in France and asked for a divorce, which Maria refused. In 1867, while still married to Maria, Berk married his mistress in a French civil ceremony, and, unbeknownst to Maria, lived with the second wife as a married couple in Paris for the next eighteen years.

Berk set up a munitions factory in France, developed a new revolving barrel machine gun, and made a second fortune in Europe. But in 1885, unexpectedly, he died. It may come as no surprise that Berk left no will; he must have known he could not have two legal widows. Before his death, neither Maria nor the second wife knew Berk was a bigamist. In the course of settlement of his estate, estimated at twelve million dollars, a legal battle ensued. Maria received the bulk of the assets; the second wife inherited nothing.

As a result of her windfall, Maria was approached with many

suggestions for memorializing her late husband, including a visit from the president of Yale University, Timothy Dwight. Dwight's proposal for the wealthy widow was to finance a private boarding school in the area of her native Salisbury, Connecticut, that would prepare young men for college—for Yale, particularly. Maria, an iron-willed Yankee, was not easily convinced. Not until the Yale president assured her that deserving boys from the surrounding communities would receive free tuition did she finally accept Dwight's suggestion. She provided the initial endowment for the school, which was founded in 1891. Her endowment also launched scholarship aid to deserving students.[66]

George Van Santvoord, the longtime Hotchkiss headmaster, universally referred to as *the Duke,* claimed that the most important concepts the school could teach were: *Be a gentleman. Be a person of character.* Breakfast at the Duke's table, after he ladled out the hot cereal, evolved into his prodding the students to debate some of his favorite topics. For example, one meal began with a discussion of *How Hot is Hell?* When the Duke's query resulted in a round of exasperated groans, the Duke chided the students, "I've heard almost every one of you complain it was *hot as hell* ... I don't even know whether they use Centigrade or Fahrenheit in hell. Now, what do you think?"[67]

Tom Brittingham arrived at Hotchkiss in September, 1938, at age fourteen. His weekly letters to his mother and younger siblings, then living in San Antonio, provide an insightful glimpse into his maturation. He described his activities and thoughts at Hotchkiss, his academic courses and grades, the books he read, the movies he saw, the athletic events, the weather, the food, and even the conversation at the dining table of the master at whose table he sat for each rotating two-week period. Every letter, usually written on a Sunday evening, began by thanking his mother for her letters and any packages he had received that week. He asked his mother what to do with old shoes after he bought new ones. Tom concluded one of his early letters, "I will stop here as there is a large water fight in progress, which I am about to join."

In December Tom wrote his first-grade brother, John, about an unanticipated holiday at Hotchkiss. He described an eight-mile hike to the Caves. The boys had to enter by crawling through a tunnel, and many of the various pathways forced them to crawl through cold water.

If they trained their flashlights on the ceiling, they could see bats. After they returned to school, they saw a movie, *Brother Rat*. "It was super," he exclaimed.[68]

Tom rejoiced in movies. Hotchkiss showed a movie every Saturday night, and the boys also saw movies in nearby Lakeville. Tom was a great fan of the Marx Brothers and Bob Hope. Among the many movies he enjoyed were *Pygmalion, Jesse James,* and *The Grand Illusion*. He called Bette Davis' acting in *Dark Victory* "superb." He liked the cartoons as much or more than the feature films. His favorite was Donald Duck. By the time he was a teenager, his sense of humor was well developed.

Tom became a voracious reader at Hotchkiss, ranging beyond the assigned works. During his first year the Civil War entranced him. He studied all the major Civil War battles and was particularly fascinated with the Battle of Fredericksburg. In addition to *John Brown's Body* and *The Red Badge of Courage,* he read Grant's *Memoirs,* the biographies of Confederate generals Jeb Stuart and Nathan Bedford Forrest, and the first volume of Carl Sandburg's *Abraham Lincoln: The Prairie Years and The War Years.*

Tom and his classmates studied many of the classics, including Shakespeare and Milton. Additional assignments included some contemporary authors. During his second year they read a novel by a new guy named Ernest Hemingway: *For Whom the Bell Tolls*.

Tom's favorite novels were *The Yearling* and *Huck Finn*, until he discovered the Russian writers during his last year at Hotchkiss. He enthusiastically wrote that Tolstoy's *Death of Ivan Ilych* was the greatest short story he ever read. When he read *Crime and Punishment* by Fyodor Dostoevsky, he reported that it was without doubt the best book he had ever read. Next he decided that *The Brothers Karamazov* was even better. Finally he wrote to his mother that Dostoevsky's novel *The Idiot* was the finest novel he had ever read. For his senior paper, the major work of his years at Hotchkiss, he wrote an analysis of the novels of Dostoyevsky.

He had only one science class at Hotchkiss, a course in physics. He never had a class in biology or chemistry.

Prep school in the East included more than academics. In November, 1938, Tom enjoyed a surprise weekend in New York City with his Aunt Margaret and Uncle Bryan Reid, from Forest Hills, Illinois. Friday evening they ate supper at the Waldorf, while listening to Benny Goodman. Then they attended the Broadway performance *Hellzapoppin*, which delighted Tom. Saturday they went to Philadelphia to see the Army-Navy football game, after which they proceeded to Wilmington, Delaware, to see Uncle Tom and his family. This experience probably influenced Tom's planning for the following fall. He suggested the Yale-Princeton game at New Haven on November eighteenth, a show Friday night before the game, another show or movie Saturday night, then either "fool around" or go to a pro football game Sunday afternoon.[69]

Hotchkiss insisted that at least four hours every day were to be spent studying outside class. Sometimes the boys actually did study four hours a day, but more often the Hotchkiss boys acted like teenage boys.

They loved to pester Dennison Fish, a music teacher, who was one of their dormitory supervisors. Fish, known among the boys as "Flip-flop Fish," had favorites. Tom, not one to kiss up to his teachers, was not on Fish's favored list. Several times Fish caught Tom with cheese, crackers, and cake, which were designated *illegal food,* in his room.

Once Tom and his roommate, Don Durgin, were kicking soccer balls around in their room, skimming them by the lamps. After lights out, they continued to fire away at each other in the dark, when Fish burst into the room. Tom, but not Durgin, got reported.[70]

Fish reported Tom often enough that Headmaster Van Santvoord sent a letter to Tom's mother about one of his *illegal food* violations.

One night Tom caught an enormous rat, undoubtedly attracted into his room by the cheese, and hung it by its tail in the hallway just outside Fish's room. Tom knocked on Fish's door, then scampered back into his own room. On another occasion he trapped a rat in a wastebasket and pounded him on the head with a math book. A few months before graduation, he found a shoe brush which exactly fit the rat hole in his room, finally victorious over the rats.

At Christmas vacation Tom's mother sent him money for the $110 round trip tickets and Pullman fare from Millerton, Connecticut, to San Antonio, Texas. On Friday evening he saw movies at Grand Central Station, gorged on a steak dinner there, and rushed to the Pennsylvania Station to catch a sleeper train to St. Louis. He then transferred to the

Texas Special of the M.K.T. (Missouri, Kansas, and Texas Railroad) for another overnight trip to San Antonio, his family's new home town.

On the return trip from San Antonio to Hotchkiss, Tom enjoyed a large poker game and met a number of pretty college girls. He wrote his mother that he would like to fly home at Easter.[71]

Tom was plagued by frequent rhinitis and sinusitis as a boy and teenager. Once when he developed a stuffy nose and slight cough he made the mistake of drinking some hot chocolate at breakfast just before going to the infirmary to be checked by Dr. Weiler, the Hotchkiss physician. When Dr. Weiler found Tom's temperature elevated, he quarantined Tom in the infirmary, called the *jug*, for four days. Tom blamed his elevated temperature on the hot chocolate and felt that his quarantine was not justified. That may have been a stimulus for his lifelong skepticism of doctors' pronouncements

The future doctor's first clinical opinion originated during a measles epidemic at Hotchkiss. Several boys were in the *jug* with measles. Tom had sat opposite one of them at the dinner table just ten minutes before the boy went to the infirmary. He wrote that the shots to prevent measles made everyone fell terrible for three days, but made the actual case of measles a little less uncomfortable.

After concluding his Lower Middle year at Hotchkiss in June, 1939, Tom visited the New York World's Fair with his friend Carter, then traveled to San Antonio briefly before summering with his mother, sister and brothers in Carpinteria, California. In September he visited Aunt Margaret and Uncle Bryan Reid in Illinois, where Uncle Bryan gave him an early Christmas gift, his own typewriter. He took a train called the *Commodore Vanderbilt* from Chicago to New York.

The German invasion of Poland on September 1, 1939, precipitating World War II, led Britain and France to declare war on Hitler's Nazi state two days later. When Tom returned to Hotchkiss, he predicted the US would be in the war within a year.

During his second year Tom took both French and German. He competed for the school newspaper, *the Hotchkiss Record*, which he described as the best extracurrricular activity in the school. Competition for the *Record* was keen and occupied most of his spare time. He made the editorial board and was a sports reporter as well.

Tom was a great fan of almost all sports. He was the only boy at Hotchkiss with a subscription to the *Sporting News*. He also asked his mother to mail him the San Antonio sports sections so he could read detailed accounts and box scores of the Texas League.

He played tennis and golfed on the Hotchkiss golf course. He also participated in soccer and ice hockey, but during the last winter he joined the Woods Squad and chopped down trees instead of playing hockey. The only sport which held no interest for Tom was basketball, which he "detested." He enjoyed skating on the local lake, and he and his friends took many hikes in the area along the Housatonic River on what is now part of the Appalachian Trail. He climbed mountains in the Berkshires, including Mt. Prospect, which offered some unscalable cliffs, and he used ropes the second time. He was an adventuresome young man who certainly did not spend all his free time studying. One day Tom and some friends hiked up a creek near Salisbury and had a great time building dams and having water fights. "If you want to have fun, there's no place like a creek," he said.[72]

He grew to admire the Duke. Tom was especially impressed one morning when the guest who was supposed to preach didn't arrive, and the Duke delivered an extemporaneous sermon. He called the Duke an "amazing man."

Tom read the *Yale News* and the *Daily Princetonian* regularly, had membership in *Book of the Month Club* and a subscription to *Time*. He bought clothes at Saks Fifth Avenue in New York. He described *Andrew Jackson* and *Science for the Citizen* as interesting and educational and started the *Odyssey* in English class. He reported that he liked oatmeal and wouldn't miss a chance for two bowls of it any time. He requested a small portable radio, reporting that the most punishment he could suffer for having a radio was a censure, no worse than having illegal food or roughhousing, and that he would use it only to listen to sports and newscasts. He expressed his feelings about himself: "I can guarantee to enjoy anything from sitting tight to taking a rocket trip to Mars."[73]

Uncle Tom, a Hotchkiss graduate himself, and Aunt Peg often visited young Tom at Hotchkiss. Their own boys would go to Hotchkiss, and Tom often took the cousins to see college football games: Yale vs. Princeton, or Army vs. Navy. He also took them to Broadway plays. Uncle Tom tried several times to interest his nephew in the world of business and finance and tried to arrange for him to spend a summer in Wilmington, Delaware. Tom declined. Uncle Tom sent Tom a book, *Seven Types of Inflation*, and articles from the financial press, trying to spark an interest in money and investing. Nothing happened. Uncle Tom, one of the most astute investors of his day and the winner of a national investment prize from *Barron's* in 1940, kept putting his money on young Tom Brittingham to enter the business world. But trying to turn his nephew into a financial man was probably the only losing proposition Uncle Tom ever undertook.

In the fall of 1940 Tom began his final year at Hotchkiss. That year T.S. Eliot wrote *East Coker*, the second of the *Four Quartets*, whose last sentence, "In my end is my beginning," would become the epitaph on Dr. Tom Brittingham's tombstone forty-six years later.

In his last year, Tom gave up the chase for grades and began his search for the truth. He was working harder and more conscientiously, and stated that he was just beginning to learn how to learn. Before he came to Hotchkiss he could write 250 words on a theme with difficulty, now he could write 10,00 words with ease. He wrote his mother that at last she was getting her money's worth for sending him to Hotchkiss.[74]

When Tom's grandfather Judge John Matthews, whom he called Otherpapa, died in the spring of 1941, Tom wrote to console his mother in Texas: "We will all have to go to face our Maker sooner or later when our trials in this world are complete … Otherpapa was a great pioneer and we must press on … For peace of mind and spirit we can only trust to God's goodness in all things."[75]

Tom had developed his religious beliefs under the watch of his mother, a devout Presbyterian. He was taught the Bible at home and in school all the way through Hotchkiss. While at Harvard Medical School five years later, he attended services at Trinity Church, an Episcopal church in Boston, and often sent copies of the sermons to his mother. These beliefs must have influenced him years later when he faced his own death from cancer with acceptance, peace, and dignity.

Before graduation that spring Tom played a round of golf with some of his classmates. His cousin, Bryan Reid, took a practice swing and hit Tom in the teeth with his driver. Tom's hands went to his mouth, he bent over in pain and bled a little. He finished that round, but during the next few weeks he made several trips by bus to Hartford for dental care. Being hit in the teeth with a golf club ended Tom Brittingham's golfing career.

Tom completed Hotchkiss ranking ninth academically in his class of ninety-six students. He was elected to the Cum Laude Society, which he called *Cum Lousy*. The Duke reported to Mrs. Brittingham that Tom was "an excellent citizen and an exceptional student." Tom had been accepted at Princeton and planned to enroll there in September, 1941.

But the impending war was increasingly on his mind. At Hotchkiss he met an exchange student from England who planned to join the Royal Air Force in September. It was an awakening to know one of his classmates would be in the war.

In May he wrote to his mother about the tough times the British were experiencing and that he predicted the United States would be in the war before the end of summer. "Britain certainly needs all the help we can give her, and it wouldn't be at all fair or sensible to let England alone try to beat Germany, Italy, Russia, and half of France."[76]

Tom had thought seriously about going to Cal Tech after Hotchkiss, but all his uncles discouraged that, advising that he go to a *regular* college like Yale or Princeton. Uncle Watt Matthews, Class of '21 at Princeton, argued that Tom should go to Princeton. In the fall of 1941, three months before the bombing of Pearl Harbor, Tom entered Princeton, with Don Durgin as his roommate. Durgin recalled that Brittingham excelled in both sciences and the humanities, but never appeared to be a grind, a term reserved for students who studied all the time. Tom never missed a sporting event.

After a year and a half, he withdrew to join the United States Army. Many years later he wrote, "What a colossal difference once one becomes hungry for knowledge ... Most effective education must be self-education, don't think the school makes too much difference."[77]

Many Princetonians were interested in Officer Candidate School (OCS), but Tom enlisted as a private. He never received a college degree. He was inducted into the Army March 4, 1943. At the time of induction he stood five feet, eleven inches tall and weighed 153 pounds. His eyes were blue and his hair was blonde.

From his Army pay he requested that $3.75 be withheld monthly to purchase Series E War Savings Bonds, with his mother as beneficiary in the event of his death. He also designated that $6.50 be withheld monthly for purchase of a $10,000 life insurance policy, also payable to his mother in event of his death. With Tom's entry into war his childhood ended.

He went through basic training at Camp Roberts in California. He wired his mother, "Setup perfect and deluxe except no town as big as Albany (Texas) within a hundred miles. Do not send shoes for present."[78] At the end of basic training he had earned qualification as a Rifle Marksman.

T.E.B. in U.S. Army, 1943

The Army enrolled him in a specialized training program in electrical engineering at the L. C. Smith College of Applied Science at Syracuse University. In view of his excellent scholastic record there he received a special insignia, a blue star on a brown background, to wear on his uniform.[79]

Brittingham participated in the invasion of the Philippines in the latter part of World War II as a member of a signal service battalion, part of Krueger's Sixth Army. The Signal Corps often preceded combat corps to establish communication. After the war he said, "Being in the army taught me not to worry about death. If I was to worry about death, that's all I would think about, I'm not going to worry about death."[80]

His duties included radio installation and repair. Sometimes he and his company had to shimmy up tall palm trees. Brittingham had fear of heights but felt if his team had to do it, so did he. At night there were total blackouts, no talking or smoking. He lay on the ground at night and thought about his life. After the war, he never took his children camping; he said he never wanted to sleep on the ground again.[81]

"I can remember what a terrible medical student I would have been had I gone straight from college to medical school," he wrote. "Having the years in the army to think about what I really wanted to do, and especially having the time for introspection in Mindanao (where we could have no light of any kind … when it got dark, you just had to lie down and think) made a phenomenal difference in my motivation."[82]

He advanced to sergeant by the time of his discharge on January 26, 1946, having spent one year, seven months, and 22 days in continental service and one year, three months, and one day in foreign service. According to his honorable discharge document, his awards included the American Theater Campaign Medal, the Asiatic Pacific Campaign Medal with two markers, the Philippine Liberation Ribbon with one marker, a Good Conduct Medal, a World War II Victory Ribbon and two overseas service markers.[83]

"He didn't like Princeton, the action was elsewhere," said Dotsy. "He had been a pampered poodle as a child, but he loved the army. He was in combat, but he never shot anyone and he was never shot at. He irritated his top sergeant, who was vulgar and insulting. He had grown up by the time he left the army.

"After discharge from the army, he lived at his mother's home in Fort Worth and completed premedical requirements at nearby Texas Christian University. He was accepted at Harvard Medical School without an undergraduate degree. At Harvard, he put TCU on his dormitory door. He thought Princeton might have sounded *'too tony.'* "[84]

6

The Most Important Decision of his Life

"I've got to get through medical school and that's all there is to it, because I know I wouldn't be very happy being anything but a doctor." T.E.B.

With the war over, in September, 1946, Tom Brittingham followed in his father's footsteps and enrolled at Harvard to study medicine. Tom described Vanderbilt Hall, the dormitory for medical students, as a rather hideous edifice. As for the rest of Harvard Medical School, he found that the five buildings of the school were beautiful Greek marble, and interconnected to save the students from walking in the rain.[85]

Of the first 128 first year students who had been selected from 4500 applicants that year, two-thirds were military veterans, and ten were women. Feung Lee, who lived in the connecting room to Tom, had been a captain in the U.S. Army. Feung Lee, who later became a prominent surgeon in Boston, taught Tom to appreciate Chinese cuisine. Tom described him as a peach of a fellow, whose habits and speech were "as American as any other citizen."[86]

When Tom arrived at Harvard, his sister, Sally, was a junior at Wellesley College, thirteen miles away, and John, his sixteen-year-old brother, was at Hotchkiss. Tom persuaded his mother to economize her correspondence by writing one letter to all three children and allowing them to distribute it. Each of his weekly letters to her still began, "Dearest Mom." In his first letter from medical school he wrote, "If I can get through this first semester, I'll do all right, but this first semester could prove tragic."[87]

There was good news as well. Uncle Tom gave Tom a new beige 1946 four-door Ford sedan. Uncle Tom gave all the Brittingham boys who served in the war a new car as a thank you. J.D. and Martha Cloud, who had worked for Lucile in Cleveland, San Antonio, and finally Fort Worth, delivered the car to Boston. J.D. served as butler, driver, and jack of all trades. Martha was the household cook. A measure of the devotion Tom held for Martha was exhibited sixteen years later. While Tom was teaching at St. Louis City Hospital, he drove from St. Louis to Fort Worth to see her every few weekends during the final stages of her illness. After the car delivery Tom took J.D. to lunch, then showed him around the medical school, including the anatomy dissecting room. "I uncovered my corpse for him," Tom wrote to his mother, "but J.D. didn't want me to unstitch him and open him up."[88]

Tom's cadaver dissection partner in the anatomy lab, Dave Kliewer, was a Marine Corps major. Kliewer, who became a lifelong friend, was the son of a poor minister. He flipped a coin with his younger brother to determine who would pay the other's way through medical school. Dave lost the toss, and became a Marine pilot. He was captured by the Japanese at Wake Island at the onset of World War II. He spent the next four years as a POW, during which time the Red Cross provided him with a copy of *Gray's Anatomy*. He read it—word for word—three times, developing a consuming passion to be a doctor. Kliewer knew more about anatomy than anyone in the class.[89]

Despite Kliewer's help, Tom struggled that first year in anatomy, and even more in histology (microscopic anatomy). He often studied until two or three a.m. and on weekends, but took a few hours out to listen to the World Series between the Boston Red Sox and the St. Louis Cardinals. He was so worried about his work that he didn't even try to get a ticket for one of the three games played in Boston.

But a life of all work was not tolerable. Tom enjoyed raucous parties in Vanderbilt Hall as well as in local honky-tonks. In October he wrote "Tonight there are going to be ten of us in my car going out to some roadhouse. We ought to have a pretty good party."[90] He continued the next week, "Had a wonderful time last night with a girl from Pine Manor.

She was a blind date, but she was a very sharp looker and an even better dancer. She also proved to be very capable with a whiskey bottle."[91]

With women now in the picture, Tom paid more attention to his appearance. He had developed considerable frontal balding. A photo of his father's head at the same age was virtually identical. Tom applied *Frances Farmer hair dope*, shampoos, and tonics to preserve his scalp hairs. He said that at least fifty of his class were also future cue-balls.

One evening he drove to Wellesley to visit his sister, Sally. When he knocked on Sally's door at Beebe Hall, Sally's roommate from Long Island, Dorothy (Dotsy) Mott, answered. She wasn't impressed. He was just her roommate's not-too-good-looking brother, and she paid him little attention. She saw him off and on at parties and dinners during the next year and a half, but nothing came of it. In fact, during that time Dotsy met and became engaged to a graduate student at Boston University.[92]

The most interesting part of his week was the Saturday morning clinic, held in one of the four Harvard teaching hospitals. Hospital clinicians lectured about real patients, and sometimes the first year students got to watch operations. Tom learned that his family's old friend, Dr. Eliot Cutler, then chief of surgery at the Peter Bent Brigham Hospital, insisted that everyone wear blue in the operating room, that even the sheets there were blue, to minimize glare on the surgeon's eyes.

Tom canceled his Christmas vacation to Texas because of his need to study. He wrote to his mother. "I've got to get through medical school and that's all there is to it, because I know I wouldn't be very happy being anything but a doctor."[93] After studying during Christmas vacation he added, "Anatomy and I are about as miscible as oil and water."[94]

He received D's in both anatomy and histology the first semester, which prompted a letter on February 18, 1947, from Dr. Dale G. Friend, Assistant Dean. Dr. Friend related that Tom's work for the first half year was not at a satisfactory level. He had made a low pass in both Anatomy and Histology. Unless his work improved greatly in the second half of the year, his continuation in the program was unlikely. Tom framed this letter and later displayed it on the wall of his office at Vanderbilt. He thought this letter represented the worst hole he had ever been in. But he also thought that getting out of it would be a great character builder, and would make him a better doctor. "If I don't," he wrote, "I don't have the stuff and would probably have been a bum doctor."[95]

After a few months he wrote to his mother that she didn't have to wonder whether or not he wanted to go on with medicine. "The longer I'm here, the more I want to be a doctor, so I will be one, whether or not I have to graduate from Podunk Medical to get an M.D."[96]

In the second term his grades improved and his mood lightened. He returned to Harvard after summer vacation to begin pathology, pharmacology, and bacteriology. "This life is a hard one, but it's an awful lot of fun," he wrote.[97]

By the third year Tom was in the top twenty-five percent of his class. That year a member of the faculty lectured on a clinical topic at 8 a.m. every weekday. These lectures were optional, but Brittingham always went to them.[98] Tom also found the lecture series on medical psychology at the Boston Psychopathic Hospital fascinating.

With his entry into clinical medicine at Harvard's teaching hospitals, Tom met immigrants, people from varied ethnic and religious backgrounds, and individuals dealing with poverty. As a result of his exposure to such diverse patients, he began to reflect on his own privileged background, "People that spend their whole lives or most of them in the circle of the well-off and 'well-bred' are greatly to be pitied."[99]

Tom saw his sister Sally at Wellesley frequently, and she borrowed his car for skiing weekends, an activity Tom did not share. On one weekend Sally and her roommate Dotsy slid into another car near White River Junction, Vermont, and severely damaged the Ford. Tom never said one critical word to his sister about the accident and had the necessary repairs done at his expense.

Subsequently, Tom's trips to see Sally at Wellesley increased. He called Dotsy Mott "a very entertaining character."[100] Tom Brittingham and Dotsy Mott had their first date in May, 1948, a month before she graduated from Wellesley. The next day he wrote his mother, "I went out last night with Dot Mott, and she really is a winner ... Sally picks her friends with exquisite taste."[101] When his mother asked why he had waited so long to ask Dotsy out, his answer was that he thought it was rather awkward to date one's sister's roommate.

Dotsy said she fell in love with Tom that night and announced to Sally that she would be marrying her brother. She broke her engagement to the graduate student at Boston University and said she never regretted her decision.[102] The couple were incurably in love and remained so until Tom's death thirty-eight years later. Tom said that his decision to marry Dotsy was the only decision in his life the correctness of which he was absolutely certain.[103]

In the latter part of May, Tom invited Dotsy to a baseball game. From the bleacher seats in right field, they watched the Boston Braves play the Brooklyn Dodgers. Six weeks later and engaged to be married, Dotsy sailed aboard the Queen Mary with a classmate on a two-month post-graduation trip to Europe. A December wedding was planned. Tom warned her, "I really sort of enjoy a hard life and I hope you do."[104]

As the lonely summer for Tom wore on, he wrote to Dotsy that emotion welled up within him, and he was so unused to having emotion of any sort, that he couldn't turn it off."[105]

During their seven month engagement, Dotsy and Tom often went to Vanderbilt Hall for parties and socializing. Dotsy said that Tom loved parties before he got so serious in medicine. She recalled the caretaker of the residence hall, a man named John, being kept drunk by the medical students so he would not enforce the visitation and curfew rules too strictly. She remembers boisterous parties with lots of alcohol, but no drugs. She said Tom was "A wild Indian with wild Indian friends" at these parties.[106]

A few weeks before his marriage, Brittingham took dancing lessons at Arthur Murray's studio in Boston. He and Dotsy practiced waltzing in the ballroom of the Biltmore Hotel in Manhattan so they would be in good form for their wedding reception.

Dotsy had grown up in Brightwater, Long Island. Her parents had a weekend home in Quogue, Long Island. She and her brother were expert sailors and owned a 24-foot sloop with a spinnaker sail, a big cockpit, and an inboard motor. As a teenager, she raced sailboats. Her father was in the raw sugar business. "Raw sugar was like oil," Dotsy said. She went to Stuart Hall, a boarding school in Staunton, Virginia, then to Wellesley, where she majored in English composition and was editor of the Wellesley student newspaper.[107]

Their wedding was on December 18, 1948, at the beginning of

Tom's Christmas vacation as a third year medical student. The site was The Church of the Heavenly Rest, an Episcopal church with gothic architecture at the corner of Fifth Avenue and 90th Street. Dotsy and her brother had gone to a school at the church from kindergarten through the second grade. Rather than a honeymoon, the newlyweds went to their apartment and "played house."[108]

Wedding of Tom and Dotsy Brittingham, December 18, 1948

Dotsy and Tom often invited a close friend, Laurens Parke (Lawrie) White, to dinner at their apartment on Beacon Street. Tom was a senior in medical school, and Lawrie, one year older, was an intern at Massachusetts General Hospital. After dinner, Tom and Lawrie spent hours talking medicine. Dotsy recalls Tom and Lawrie questioning and challenging conventional medical wisdom. She believed that Tom's opportunity to critically discuss medical issues—first with Lawrie White in Boston and later with David Rogers in New York—fertilized the striking unconventionality of his medical thoughts. She thought these late night talks changed Tom from a conventional thinker to a challenger who doubted conventions and saw flaws in the system.[109]

During the following summer, Dotsy and Tom asked John Burnum, who had joined Tom's third year class at Harvard as a transfer student from the University of Alabama, to spend a weekend on Long Island at the home of Dotsy's parents. Burnum recalls they sailed to Fire Island on a thirty-foot sailboat, Dotsy doing most of the sailing, all the while serving white wine and lobster to her husband and their guests.

Tom's most important decision during his final year at Harvard was where to acquire internship and residency training. He was accepted for medical internship positions at the Peter Bent Brigham Hospital in Boston and at New York Hospital-Cornell Medical College. He was tempted to follow in his father's footsteps at the Brigham. However, he decided to go to New York Hospital. Dotsy's parents lived in New York, and he knew that Dotsy would appreciate their proximity while he was so occupied by hospital work.

At New York Hospital Tom would meet Dr. David Rogers, who would have a major influence on Tom's future; several years later, they would reunite at Vanderbilt.

7

White Suit Doctors

"The clearest memories I have of Tom are his frequent midnight appearances on the medical floor, when, at this quiet time, he would sit down and doggedly review every last bit of data on a problem case, including a thorough review of multi-inch thick previous records, gleaning every possible last bit of information about that patient, and often coming up with pertinent and overlooked data."—Carl Wierum, M.D.

Dr. Tom Brittingham arrived at New York Hospital in the summer of 1950 to begin his postgraduate medical education, residency training. In the making of a doctor, the residency represents the dominant formative influence. During the resident years a doctor not only acquires the knowledge and skills needed in his specialty but also develops habits and attitudes that last a professional lifetime.

Although clinical training had its roots in an apprenticeship system, the modern residency system arose from the revolution in scientific medicine in the late nineteenth century. At that time, German universities were the international leaders of scientific medicine and research. The first formal residency program in America began at Johns Hopkins Hospital in Baltimore in 1889. The program developed at Hopkins ultimately served as a prototype for medical training programs throughout the United States. This was true for the New York Hospital-Cornell Medical College, where Dr. Brittingham trained in internal medicine; the Barnes Hospital-Washington University School of Medicine, where he studied the specialty of hematology; and Vanderbilt University Hospital, where he achieved maturity in his career as a physician-teacher.

The Hopkins medical faculty, adhering to the German model, felt that advanced clinical training should take place in a specialized teaching hospital and should represent true graduate education rather than just vocational training.[110] The most important factor affecting the quality of residency training was the standard of patient care maintained in the environment. Aware that medical knowledge was rapidly proliferating, the Hopkins faculty was skeptical of traditional medical authority. These were attitudes shared by T.E.B. throughout his career. Medicine at Johns Hopkins was more absorbed than taught.[111] Young doctors attended interactive rounds and conferences that evaluated the evidence underlying medical assertions. For example, Morbidity and Mortality Conferences at Hopkins explored errors of medical practice and offered alternative treatments. These were the model for Dr. Brittingham's dramatic Death Conferences at St. Louis City Hospital and Vanderbilt.

The Hopkins house officers experienced total immersion in the institutional culture. They all worked, ate, and lived in the hospital. The hospital was more than an institution, it was a way of life. Typically the interns and residents were *on call* (admitting new patients and managing unforeseen problems in patients already hospitalized) every other night. This system also prevailed at Vanderbilt until the late 1970's, when the faculty changed it, over T.E.B.'s objections, to every third or fourth night calls, because of concern about the effects of resident fatigue. T.E.B. said then, with his characteristic chuckle, "The only problem with being on call every other night is that you miss half of the good patients." This had been the attitude at Hopkins since the 1890's.

Though complaints about poor food, inadequate living quarters, and the lack of fresh air and exercise were frequent, many residents seemed almost oblivious to the long hours. In these circumstances, illness among house officers was common, with pulmonary tuberculosis a particularly notorious threat.

The success of the Johns Hopkins residency system was due to the efforts of the energetic and beloved Physician-in-Chief, Dr. William Osler. Osler's interactions with patients demonstrated the same concern for thoroughness and attention to detail which Dr. Tom Brittingham exhibited later. While they were still bachelors, Osler and Dr. William Halsted, the chief of surgery, also lived at the hospital. Osler did not marry until he was forty-two. He wrote *The Principles and Practice of Medicine*, the first great textbook of modern medicine, in his living quarters at the Johns Hopkins Hospital in 1892.

This was the golden era of physical examination, which clinicians perfected to a degree scarcely imaginable by today's physicians. Instead of high-cost testing, which was not yet available, a doctor's bedside evaluation was the key to diagnosis. There was an unwritten contract for patients unable to pay—in return for free medical care, they would be *teaching patients*. Commercialism was antithetical at teaching institutions at that time; professional ideals dominated the culture.[112] To succeed, doctors needed to focus nearly totally on medicine, to the exclusion of family, friends, hobbies, and a balanced life. The surgeon Charles W. Mayo said, "I'd never advise a girl to marry a doctor if he is a good one. She will be in for a lot of loneliness."[113]

Brittingham planned to enter the field of internal medicine, his father's specialty, and opted for a *straight medicine* internship. He and Dotsy moved into an apartment on York Avenue, in a building owned by and across the street from New York Hospital. The previous custom regarding marriage of house officers had changed; many residents were married and even had children.

Tom and Dotsy already had a young son, John, who was born in Boston. Tom described John as "So terrific, so perfect, such an overwhelming pleasure. We're so darned lucky to have one like him."[114] During the next four years in New York, the couple would have three more offspring, Susan, Harold, and Margaret. Dotsy's parents and her Aunt Poosie provided invaluable assistance to the growing young family. Tom's life would be hectic, starting with a year as an intern, then two years as an assistant resident, and a fourth year as the chief resident in medicine.

The first patient that Tom admitted to New York Hospital as an intern gave him an immediate connection to Dr. David Rogers, who would play a pivotal role in his medical career. The patient, Corahelen Baxter, was Dr. Rogers' mother-in-law. Mrs. Baxter had sustained a massive stroke and died later that day. David Rogers and his wife, Corky, had recently moved to New York so that he could pursue research interests in infectious diseases at the Rockefeller Institute.

After the funeral, the Brittinghams and the Rogers' arranged to have dinner so Dotsy and Corky could renew their friendship. They had been hallmates at Wellesley College. In fact, when Corky and David became engaged, Dotsy and her roommate Sally Brittingham congratulated them with a bottle of wine they had buried for safekeeping.[115] Later, Dotsy hosted the two couples for a sailing weekend at her parents' home on Long Island. Thereafter, the two families, both with two small children, often met socially. Within a few months, Rogers and T.E.B. had become tight-knit friends and would remain so for the rest of their lives. Theirs was "a pure friendship, everything a friendship should be," said Dotsy.[116] During their social evenings, T.E.B. and Rogers talked medicine constantly.

David Rogers, two years Tom's senior, had graduated first in his class from Cornell University Medical College and had spent the next two years as an intern and assistant resident at Johns Hopkins Hospital, which was considered the most desirable training program in the country. At New York Hospital he was chief resident in medicine when Brittingham was a first year assistant resident. The miracle drug, penicillin, had recently been introduced for civilian use, but many staphylococcal infections were resistant to penicillin and were a scourge on the medical and surgical wards. Rogers' research in this serious and perplexing affliction cast him at the center of one of medicine's most difficult challenges at that time.

Tom Brittingham and David Rogers had many shared attitudes and opinions: Both had an intense distrust of conventional medical authority; both viewed medicine as a service profession, not as the route to financial success; both considered the patient and his welfare most worthy of attention; both concentrated on the basic medical skills of listening to the patient and conducting a thorough physical examination, rather than on laboratory tests, x-rays, or the newest technology; and both loved to teach at the bedside. But Rogers was different from Brittingham in one respect—he knew early that he wanted a career in academic medicine—he wanted to be a departmental chairman.

David was the son of Carl Rogers, one of the most eminent psychologists of the 20th century, among clinicians second only to Sigmund Freud.[117] Carl Rogers was famous for his psychotherapy research, especially his unique understanding of personality and human relationships. His best-known book, *On Becoming a Person*, discussed his *client-centered* approach and the development of the *concept of self*. "This process of the good life is not, I am convinced, a life for the faint-hearted. It involves the stretching and growing of becoming more and more of one's potentialities. It involves the courage to be. It means launching oneself fully into the stream of life."[118] Carl Rogers's description of a rich full life would be an apt characterization of his son, David.

After a year of internship, the Cornell faculty encouraged the medical residents to explore subspecialties during their next two years. T.E.B. did a six month rotation in neurology under Dr. Harold Wolf, whom he considered his best teacher and hero. Dr. Wolf sometimes invited the Brittinghams to his home in Riverdale, an opulent, hilly suburb in the Bronx. Dotsy remembers Wolf once greeting them half-way up the seventy-two steps leading to his front door. T.E.B. also spent six months on the hematology service, the study of diseases of the blood. He ultimately chose the latter subspecialty over neurology.

Dr. Carl Wierum was an intern in 1951-52, one year behind Tom. Wierum recalls T.E.B.'s extreme assiduousness and attention to detail and told me, "What I remember is his absolute desire to get all the information possible, and I'm sure some of that rubbed off on me. My relationship to him was more of an intern-resident than a social friend. I don't know whether he had time for friends."[119]

As Brittingham was finishing his second year of residency and preparing to become chief resident in medicine, he received a letter of advice from his friend David Rogers. Rogers told T.E.B. that his modesty was becoming, but he carried it to a fault. He warned T.E.B. that in his own experience as chief resident he found himself supervising junior residents who knew as much medicine as he did, and that T.E.B. in that position did not have to be the end-all authority in each area. He recommended that T.E.B. strive to create a happy working atmosphere in which all his junior residents could shine without treading on each other's toes, an atmosphere in which everyone could learn from everybody else. And he concluded by reassuring T.E.B. that he was indeed the man for the job.[120]

When T.E.B. was chief medical resident, one of the junior residents described him as the most consistently serious and concerned man he had ever met—always there, knowing every sick patient, running down the halls with his head thrust forward, always aware of the dangers and pain facing the patients. This junior resident was amazed to learn later that T.E.B. had a life outside the hospital.

An intern remembered the feeling among the intern class that no amount of effort on their part would prevent T.E.B. from uncovering something important about their patients. But T.E.B.'s presence as chief resident created emotional security among the interns and residents he supervised, making it possible for them to work harder, to do their work better, and to be happy doing so.[121]

In February of his year as chief resident (1953-1954), Tom received the following presents for his thirtieth birthday: underwear; tie; tennis racquet; a new book by E.B. White, who later co-authored the writers' manual *The Elements of Style*; books about Botticelli and Monet; two jazz records; and an antique magazine rack. He had recently read the newly published 661-page book of art criticism by André Malraux, *The Voices of Silence*. Tom called it the toughest reading he'd ever attempted. Even with the presence of *Webster's Collegiate Dictionary* and *The Reader's Encyclopedia* constantly at hand, he said he had understood very little of the book.

At the completion of his year as chief resident, T.E.B. felt satisfied. Dr. David Barr, Chairman of Medicine at New York Hospital-Cornell Medical Center, wrote a letter of recommendation to Dr. Carl Moore, his friend and former colleague at Washington University in St. Louis. In that letter Dr. Barr, not a man to exaggerate, described Tom Brittingham as *solid gold*. Dr. Barr assured Dr. Moore that he would like Tom personally and would find him a very able hematology research fellow. Dr. Moore replied that he would be delighted to accept Tom. It was as easy as that.

Tom received a two-year fellowship, funded by the United States Public Health Service, with a stipend of $4800 for his first year as a research fellow in hematology.

8

Self-Experimentation at Barnes Hospital

"Work brings me peace of mind, vacations restlessness."—T.E.B.

Dr. Thomas E. Brittingham II was the chief resident on the Cornell Medical Service of New York Hospital when he came to St. Louis on Thanksgiving weekend of 1953 to interview for a fellowship in hematology at Washington University. His main interviewer was Dr. Carl V. Moore, the silver-haired Chairman of the Department of Internal Medicine and Chief of the Division of Hematology. To Brittingham's great surprise, he found that Dr. Moore was a patient in Barnes Hospital. Dr. Moore's nose was oozing blood. Nurses interrupted frequently to check his blood pressure and pulse rate, but Moore went ahead with Brittingham's bedside interview. He sat propped up by pillows to prevent movement of his head and seemed unruffled, as if he were in his own office.

Dr. Moore had become a patient as a result of a self-experiment to study the cause of idiopathic thrombocytopenic purpura, or ITP— a blood disorder characterized by a lack of platelets, a type of blood cell essential to clotting. "Idiopathic" is a Greek word which means "of unknown cause." "Thrombocyte" is the medical name for a platelet, and the suffix "-penia" denotes a deficiency. "Purpura" is from the Latin word for purple; it describes the black and blue marks caused by the leakage of blood into the skin. Occasionally ITP can result in death from internal bleeding. Dr. Moore's nose had bled so profusely that his doctors feared that bleeding could occur in his brain, resulting in a stroke.

Moore's scientific reputation was based in part on experiments he had conducted on himself, including the swallowing of radioactive iron to study the way red blood cells used iron. His personality captivated Brittingham, and when Moore offered the fellowship, T.E.B. accepted without hesitation.

Moore, the son of a St. Louis policeman of modest means, was scrupulously honest, even though he was known to be a fierce poker player. An excellent clinician, he led hematology rounds at Barnes Hospital three times a week with a following of as many as thirty faculty, fellows, residents, and students. Brittingham experienced Moore's mild-mannered formality at the bedside. Dr. Moore stood at the right side of the patient. The resident or fellow presenting the case stood at the patient's left and handed Dr. Moore the chart. After presenting the history and physical findings from his memory, the trainee might refer to a few notes for laboratory data. Dr. Moore listened attentively to the presentation, then turned to the patient for a few additional questions. He often demonstrated physical findings; he was an expert in palpating a spleen.

Moore's patient-centered grand rounds also inspired Brittingham. After introducing the subject of the day, Dr. Moore walked to the anteroom adjacent to the amphitheater, greeted the patient in a wheelchair, and wheeled the patient before the audience. Moore stood with his hands on the handles of the wheelchair, symbolically connecting with the patient, and recited the history and physical findings from memory. During the discussion he kept the patient with him, emphasizing that the conference concerned a real human being with a real problem.

While Carl Moore encouraged excellence in clinical care in his department and division, he was even more intent on building a research program. He aspired to the Nobel Prize, if not for himself at least for someone in his department. He wished to emulate Carl and Gertie Cori, his biochemist friends at Washington University, who had won the Nobel Prize in 1947 for their work in carbohydrate metabolism. Gertie Cori suffered from myelofibrosis, a disease of the bone marrow, and was Carl Moore's patient.[122]

T.E.B.

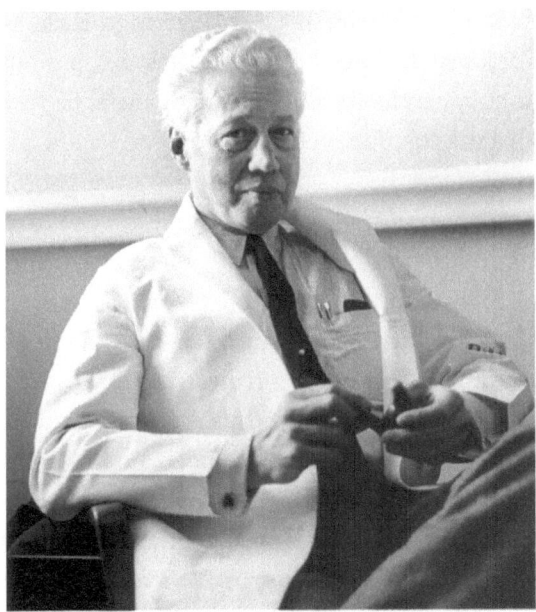

Dr. Carl Moore

Dr. Brittingham found Washington University to be a thriving clinical and research institution. Ernst Wynder, a third year medical student, had just published his epidemiological observations identifying cigarette smoking as a contributory cause of lung cancer. His co-author on the publication was Dr. Evarts A. Graham, chief of surgery, who had performed the first successful pneumonectomy (resection of the lung) as a cure for lung cancer, and who later developed oral cholecystography, the first procedure to image the gallbladder and detect gallstones. Washington University enjoyed the support of several local philanthropists in bustling post-war St. Louis.

T.E.B. heard many stories of the exploits of Dr. William Harrington, one of the hematology fellows a few years previously. Moore had encouraged Harrington to devise experiments to learn more about ITP. Harrington decided to take plasma from a patient with ITP, inject it into himself, and then measure his platelet count. If, as he suspected, the plasma contained something that destroyed the platelets, his own platelet count would decrease rapidly. It was a different era, before establishment of institutional review boards (IRB's), whose mission was to protect the welfare of human research subjects. There were no informed consent forms to sign.

Before the experiment, Harrington's platelet count was normal. As a pint of the patient's blood trickled into his vein, Harrington sat quietly in a chair, reading a medical journal. Three hours later he lost consciousness and had a grand mal seizure. When he regained consciousness, he resumed counting his platelets with a microscope. Very few were seen. He was so enthusiastic about the experiment that he was oblivious to any danger to himself.

The next morning Dr. Moore learned of the experiment. His technician confirmed that Harrington's platelet count was still very low. Dr. Moore, who had much more experience in treating ITP than Harrington, admitted him to the hospital. By that time he had black and blue spots on his legs, bleeding from the gums, and some blood in his stools. He spent a week in the hospital, as his platelet count slowly returned to normal. The experiment had been a success, proving that ITP was due to a factor in the blood that destroyed platelets.

Using a lower dose of the patient's plasma, Harrington repeated his self-experiment thirty-five times during the next few years. Virtually everyone in the department—Dr. Moore, colleagues, technicians, and secretaries—participated in the experiment at one time or another. Further research showed that ITP resulted from an immunologic abnormality—the body made antibodies which destroyed its own platelets—a condition now called auto-immune disease. Further refinement to Harrington's research led to a Nobel Prize in 1980, won by Professor Jean Dausset, an immunologist in Paris.[123]

Dr. William Harrington

No one knows how many self-experimenters there have been over the years, nor how many there are today. Harrington upheld a tradition of self-experimentation at Washington University that had begun before him and would continue long after.

When Brittingham began his hematology fellowship in September, 1954, the chief of the division, the faculty, the fellows, and the residents all functioned as experimental animals. Harrington was the boldest, but he would soon meet his match in Brittingham. One of T.E.B.'s first self-experiments involved removing two units of blood from himself and then bounding quickly, two steps at a time, up fourteen flights of stairs, to measure his exercise tolerance.

Dr. Moore assigned T.E.B. a project: to find out if white blood cells had other properties beyond their role in fighting infection. Moore and his colleagues were curious why some people had reactions to blood transfusions, even when the red cell blood types were perfectly matched. They suspected that these reactions were due to white blood cell antibodies. Brittingham fashioned a laboratory test to detect white cell antibodies. He noted that the level of white cell antibodies was highest in patients with hematological disorders who had received hundreds of blood transfusions. It was plausible that white cell antibodies, rather than red blood cell type, might be the cause of serious, sometimes fatal, transfusion reactions in these patients.

Most mornings T.E.B. arrived in the laboratory at 7 a.m., and at least four evenings in the week he worked until midnight. He wrote to Dotsy, who was visiting in Texas, that he no longer felt tense about research. He stated that at the end of two years, he would go into private practice, wherever she wanted to go, working only the hours she wanted, with plenty of time for reading at home.[124] The Brittingham family lived in St. Louis for nine years.

Brittingham designed a series of experiments which involved infusing into himself blood from patients with leukemia. He sought to answer two questions: Would a normal individual produce antibodies to the injected leukemic white blood cells? And could the cancer (leukemia) pass from one person to another?

Many scientists believed that cancer was a communicable disease. The ideal way to discover whether transmission of leukemia between people could occur would be the injection of leukemic blood into healthy individuals. But no researcher dared do it. Brittingham didn't think the risks of such an experiment were great, but he felt, for ethical reasons, the investigator should first do the experiment on himself. At the time, he was thirty years old, married, the father of four children, and his wife was pregnant again.

Dr. Moore voiced no objections, but Harrington was skeptical. He warned Brittingham he was taking the risk of getting leukemia. Brittingham responded that he would proceed with the experiment, with or without Harrington's help. Harrington shared some of Brittingham's insatiable curiosity regarding the project and understood the young investigator's willingness to put his life at risk for an experiment he believed in. Brittingham, like Harrington and Moore, was anxious to advance science. One of his oldest professional colleagues and friends, Dr. Seymour Reichlin, regarded the self-experimentation as Tom's way of serving humanity.[125]

After a year in St. Louis, T.E.B.'s research was going well. He loved his work, even during the hot, humid St. Louis summers. Since the hospital was not air-conditioned, he tried to avoid the air conditioner at home in order to build up heat tolerance. In this way he could keep working hard even on the hottest days. He installed a horizontal bar in his back yard to chin himself, and the children used it for acrobatics. He joked about Chico, a rampaging monkey in the lab, who unlocked her cage twice and ingested some research data.

Brittingham proceeded to inject white blood cells into his body from a patient with chronic myelogenous leukemia (CML). The patient's white blood cell count was approximately fifty times normal. T.E.B. then tested his own blood, first to to see whether he had produced antibodies to the injected white blood cells, and second to determine whether he had developed leukemia.

In a series of experiments, Harrington withdrew 100 millimeters (ml) of blood—enough to fill two large syringes—from the patient with CML. He then injected the entire amount into Brittingham's arm vein. By the sixth injection, Brittingham was a human antibody factory, producing abundant antibodies against the leukemic patient's white blood cells. After the seventh injection he experienced a chill and fever. Yet even during the lengthy hard-shaking chill he continued to test his own blood samples under his microscope. The observations supported his hypothesis that normal persons could develop antibodies against foreign white blood cells.

Subsequently Brittingham undertook experiments in which he injected himself with the blood of ten different patients, each with a different blood disorder. With the tenth patient, Brittingham wanted to transfuse himself with 250 ml of the patient's blood. But Harrington knew that this patient had very strong white blood cell antibodies, and the blood might produce a fatal transfusion reaction. He persuaded Brittingham to reduce the amount to 50 ml. Later Brittingham acknowledged that Harrington had saved his life by urging him to use the smaller dose.

On June 21, 1956, Brittingham received the tenth injection. He immediately felt faint. Fifteen minutes later he was profoundly weak, and pale. He vomited and had diarrhea and shaking chills. His temperature increased markedly, and his blood pressure fell to dangerously low levels. He was breathing sixty times per minute and was still short of breath. His skin turned blue. His lungs filled with fluid, a condition called pulmonary edema. His electrocardiogram and chest X-ray looked frightening. His colleagues, afraid that he would die, admitted him to Barnes Hospital, the same hospital where Drs. Moore and Harrington had required care a few years earlier.

That evening, Brittingham's wife, Dotsy, arrived at the hospital. When Tom had not called that he would be late for dinner, she had become concerned. She was outraged with Dr. Moore for allowing this type of experimentation in his laboratory.[126]

Treatment in the hospital with intravenous fluids, hydrocortisone, and oxygen worked, and by midnight Brittingham was out of danger. He went home the next day. Weakness persisted, however, and in three months he developed jaundice and hepatitis B as a result of the near-fatal transfusion. Thereafter he could not tolerate alcohol, though he had never been a heavy drinker.

Even as he was recuperating, Brittingham began another set of self-experiments—to test his most ambitious hypothesis—that injection of white blood cell antibodies into patients with leukemia might suppress the cancer. As a pilot test, he carried out the experiment on himself. Again,

Harrington repeatedly injected Brittingham with leukemic white blood cells to restimulate his production of white cell antibodies—twenty-three times over an eight-month period. Brittingham became sick after most of the injections, the severity of his reaction generally correlating with the number of white blood cells transfused.

Blood removed from him was separated into a fraction containing the antibodies. Brittingham injected this fraction—the hoped-for antiserum to leukemia—into the same woman with chronic myelogenous leukemia who previously had donated her leukemic cells to him. The antiserum decreased her white blood cell count, but the effect was transient. It worked for about two months by slightly increasing the rate of destruction of white blood cells, but it did not decrease their massive over-production in her bone marrow. Brittingham's dream of developing an effective anti-leukemic treatment ended. He did not report or publish this study or many others he did, because he disliked writing scientific papers. But his work was a pioneering effort in cancer immunotherapy, a concept very much alive today.

While Dotsy and the children spent spring vacation in 1957 visiting Uncle Watt at Lambshead Ranch, Tom stayed at work. He became restless on vacations, but work brought him peace of mind. He had already recognized disturbing changes in the medical profession and thought that doctors were less dedicated than he expected them to be.[127]

Dr. Moore recommended Brittingham to head the Section of Hematology at the University of Florida, but T.E.B. declined the invitation on the basis that he did not consider himself a competent enough investigator to hold such a position. Dr. Moore wrote to Dr. Samuel Martin, Professor of Medicine at Florida, regarding T.E.B.'s decision, "As I told you over the telephone, he is a stubborn cuss, even though a very pleasant one."[128]

In 1959 Dr. Moore put Brittingham in charge of the Washington University medical service at St. Louis City Hospital (see next chapter). Despite the responsibilities of directing the teaching program there, Brittingham's self-experiments— taking blood from patients with leukemia and injecting it into himself—continued, at least for several months. He experienced the typical reactions of fever and malaise from the foreign white blood cells and antibodies.

T.E.B.

For several years T.E.B. performed frequent checks of his blood count, but there was never a sign of leukemia. Before his death in Texas in 1986, the result of kidney cancer, he arranged for an autopsy on himself. The purpose of the autopsy was to relate the absence of leukemia in his body to Dr. Lawrence K. Altman, the author of the book, *Who Goes First? The Story of Self-Experimentation in Medicine*, the book which supplied much of the material in this chapter.[129]

9

St. Louis City Hospital

"At City Hospital, all hands were on patients ... T.E.B. loved it."—Dr. Carl Mitchell

When Dr. Brittingham accepted the job as Chief of the Washington University Medical Service at St. Louis City Hospital, he wanted the institution to become to St. Louis what Boston City Hospital was to Boston. He was a one man show, the only full-time faculty physician on the service.

Carl Mitchell and I were third-year medical students in the fall of 1959, during Brittingham's first year at City Hospital. Each of us spent six weeks on T.E.B.'s service there. Also, Carl chose City Hospital for his internship. In his extensive journal of his internship and a taped interview 56 years later, he said that *curious* was the word to describe Dr. Brittingham—curious about each patient, the patient's family, and the patient's home situation. He instilled curiosity in the medical students, interns, and residents. He approached each patient as an important problem to be solved. Much of the information in this chapter is reproduced, with Carl's permission, from his journal.

St. Louis City Hospital, just south of downtown, was surrounded by high-rise public housing. The area was rife with poverty, homelessness, alcoholism, and crime. In the hospital staff parking lot the doctors and nurses left their parked cars unlocked, the glove compartments open and all contents visible, to prevent smashed windows from theft attempts.

The hospital was founded in 1845 in response to a cholera outbreak. It was rebuilt twice, having been destroyed first by a fire and then by a tornado. The red brick and limestone Georgian revival style building had high ceilings and elongated windows. Each hospital ward accommodated thirty patients, both male and female, with fifteen beds on each side, separated by curtains. The nurses kept the curtains around the beds open as often as possible, so they could observe all the patients from the nurses' station. In summer, the south-facing wards were ovens. In winter, patients shivered under piles of threadbare blankets. Most patients were critically ill and admitted via the emergency room between 9 p.m. and 3 a.m. If a patient could walk, talk, and eat, he probably didn't require inpatient services. The house staff suffered from chronic fatigue.

Just before graduation from medical school, Carl contracted pulmonary tuberculosis, possibly from doing mouth to mouth resuscitation on a homeless, intoxicated patient. The infection delayed the beginning of his internship by several months. Effective drug treatment for tuberculosis had only recently become available. T.E.B. visited Carl frequently in his hospital isolation room, the only faculty member to do so. During subsequent months of isolation at his apartment, T.E.B. visited three times, once to visit solely with his wife, Mona, "just to see how she was doing."[130]

The lack of specialty medical services forced the young doctors at City Hospital into care beyond their competence. "A numb, proud spirit of medical care and learning enveloped us," said Carl. "We would bitch, think of cutting corners, and pray that our marriages would last."[131]

After his first week, T.E.B. was working at least twelve hours daily. Dr. Moore came to his first medical conference as a show of support. T.E.B. was shocked that most of the patients had "terrible apathy towards their own health." He complained that the food in the cafeteria was high in fat. He tried to get the house staff and students to wear ties, so that his friends at Barnes would not refer to his service as "The City Dump."[132] These were the days before cell phones or pagers. To summon physicians, hospital telephone operators launched overhead pages. Intensive care units, cardiac care units, rehabilitation facilities, and hospice care did not exist. For imaging, doctors used plain X-rays of the chest, skull, abdomen, and bones, X-rays of the gastrointestinal tract after swallowing or rectal insertion of radiopaque barium, and X-rays of the kidneys after the injection of soluble radiopaque dyes. X-ray imaging of the brain and spinal cord required injection of a radiopaque dye into the cerebrospinal fluid by lumbar puncture. Coronary arteriograms and flexible endoscopy were still being developed. CT scans, MRI's, and diagnostic ultrasound were unimagined, a decade or more away. Hemodialysis was not yet available to extend the lives of patients with kidney failure. Biochemistry, pharmacology, and radiology were just beginning to expand, after having been almost unfunded during the preceding thirty years of depression and war. Since rapid and precise diagnosis was often elusive, careful history and physical examination were all-important. T.E.B. loved it.

The first medical term I learned at City Hospital was "workup," the word for gathering information about a patient and examining him. Dr. Brittingham taught us how to workup a patient.

The medical students and interns were responsible for all blood draws and administration of IV's. They had to finish these chores, which they called "scut work," before morning rounds began at 7:30. T.E.B. was never late, unless he was still typing his own patient evaluation notes with his battered portable Smith-Corona typewriter. Steel needles for IV's were large, and IV fluids routinely damaged veins. Cardiopulmonary resuscitation (CPR) was usually a one-man job and was more of a learning exercise than an effective procedure. During nights and weekends the medical students and interns had to type the blood used for transfusions, interpret their own cultures in the bacteriology lab, and take their patients to x-ray or surgery, if they could find a gurney somewhere. Nurses were excellent but limited in number. The nurses were true medical missionaries, mostly young, and, unsuspected by them, on their way to burnout.

Mortality was frequent. Post-mortem examinations were an important part of our education, and acquiring consent for autopsy was another obligation of the medical student and intern. We competed against each other for the highest rate in gaining autopsy consent, succeeding about 50% of the time. At the weekly death conferences, T.E.B. discussed a case of his choice, often selected because it was mishandled. The conferences offered intense scrutiny of the case and contrasted what the clinicians had thought during their evaluation versus the actual facts demonstrated by the autopsy. T.E.B. was capable of indignation when he thought critical mistakes had been made, but no individual was humiliated or shamed unless there was neglect of duty. Ignorance was correctable. Many medical students maintained that the death conferences at City Hospital were the best lessons they ever experienced.

The outpatient clinics consisted of drapery-enclosed cubicles for patient evaluations. Patients and doctors could hear everything in the adjoining cubicle. Once Carl heard T.E.B. talking to a grandmother who was trying to hold her family together. She felt overburdened and was losing hope. Carl said that he distinctly heard T.E.B. say, "I'm going to have the nurse give you a B-12 shot. This should help a lot." Later Carl asked if the patient had B-12 deficiency. T.E.B. replied, "No, she needed help when none was available. It's not wrong to prescribe a placebo as long as you honestly feel that it will help and will do no harm."[133]

For the interns, a shift in the emergency room was considered soft duty compared to the inpatient wards. They learned triage and how to make rapid decisions. Sometimes they saw eighty patients in eight hours. Triage often meant immediate admission to the hospital wards, some with a diagnosis of "I know he's sick, he's yours now." The police brought many patients to the ER. Some were intoxicated, irrational, angry, and combative. Often the nurses and police had to teach the young doctors what to do while the patients cursed, kicked, and thrashed. A helpful policeman might tell a doctor to look the other way to avoid seeing what happened.

Carl said that vulgar young punks were a special irritation. One evening everyone in the ER was upset by a cursing, belligerent man. Carl asked him to calm down. With a second warning, Carl added that if the rude, loud behavior continued, he would kick his ass out of the building. The man resumed shouting vulgarities and racial slurs. No police were around. "My anger went to my legs," Carl said, "which kicked his ass off the chair and out the ER door." I asked Carl what T.E.B. said about that. "Nothing, but I received a standing ovation from the nurses, who said, 'Now there's a doctor who cares.'"[134]

Women patients were often victims of sexual abuse. Social services, rape or crisis centers, and hot lines did not yet exist. Counseling of patients was limited to "Don't drink so much, don't eat too much salt." But in the midst of poverty, social chaos, and violence, T.E.B. demonstrated his personal involvement in the life of each patient, an approach that held scant appeal or importance for most academic physicians at the time.

T.E.B. frequently conducted bull sessions at lunch. These talks provided insight into the complexity of his apolitical approach. T.E.B. had respect for all persons and their challenges. He marveled at the strengths of the poor and impaired. Once he mentioned that the government should do more for their health, then halted abruptly. He must have realized that he was overstepping some boundary in his influence with young doctors. After two years at City Hospital, T.E.B.'s political views had evolved. He felt that his contact with the the underprivileged was turning him into a political left-winger. "I shall have to keep my mouth closed at family gatherings," he wrote to his mother.[135]

Carl Mitchell said he saw T.E.B. angry only once. After morning rounds, T.E.B. asked a medical student about the results of a urine test entered on the lab sheet. The student replied that it was OK. T.E.B. said that he suspected diabetes and had tested the urine himself—it showed large quantities of sugar. The student, stunned, had been caught dry-labbing. (Dry-labbing was a slang term for recording results of lab tests that actually had not been performed). T.E.B. was furious. He had seen the student leaving early in the afternoon.[136]

T.E.B. did have some troubling experiences at City Hospital. As the internship year of 1960-61 was ending, he noted a decrease in the quality of the interns' work. The problem for T.E.B. arose in deciding how to handle poor performance. He thought people functioned better with praise and encouragement, which was his usual modus operandi; yet to condone inferiority just encouraged more of the same. For T.E.B., one of the unpleasant consequences of a leader's position was that he must not accept poor performance. His ambivalence about leadership would not end with this episode.

Dotsy took the children to her parents' beach home in Long Island for several weeks during the summers. T.E.B. was lonely, buried himself in work, prepared frozen dinners for himself at home, and sometimes ate breakfast at the Toddle House. He read books in the evenings, but skipped the opera and movies, which he felt should be enjoyed together. He went to church on Sundays, cared for his children's pets, and wrote handwritten letters to each of them, joking about his miserable penmanship. He wrote to Dotsy, "Delighted Quogue is so much fun for them ... How little they have ever had from their father, how much my father always gave."[137]

T.E.B.

Tom and Dotsy Brittingham had a date for lunch together in the park near the hospital three times a week. They had an active social life in St. Louis. They especially enjoyed dinners with their many friends, including Seymour and Ellen Reichlin and Hugh and Alice Chaplin. The Chaplins were next door neighbors, and Hugh Chaplin was also a co-investigator with T.E.B. in the hematology department.

In late fall of 1961, the shortage of medical residents at City prompted T.E.B. to take call as a resident and sleep at the hospital every fourth night. This extra work led to cancellation of the family's annual Christmas trip to Lambshead Ranch and persisted for six months, until Dotsy put her foot down.

Dr. David Rogers. T.E.B.'s close friend during house staff days at New York Hospital, called T.E.B. in early 1960 and invited him to join the medical faculty at Vanderbilt. T.E.B. declined, as he was just settling into his new role at St. Louis City Hospital. However, by spring of 1962, he acknowledged his growing frustration with the lack of support from Washington University for his beloved City Hospital. After Brittingham discussed moving to Vanderbilt with Rogers and with Dr. Moore, Moore took action to see what could be done to keep T.E.B. in St. Louis. He interceded with the Dean and with the City Hospital Administration to raise Brittingham's salary from $15,000 to $18,000 annually. But salary was not the issue.

T.E.B. held a fervent belief that patients at a city hospital, rather than private patients, were critically important for medical education. He convinced himself that Washington University and most other medical schools did not share his beliefs. Furthermore, he believed that the faculty at Washington University, devoted to research, viewed patient care and teaching as chores beneath their dignity. He thought that only a rare man was a productive investigator, and it was foolish to have attached so much value and prestige to research, to the exclusion of teaching and taking care of sick people.[138]

During his time at City Hospital, Brittingham became increasingly cynical about medical meetings, medical societies, and medical research, an attitude which would impede his success as a typical medical academician. He maintained that most medical meetings were not conducted in the interest of education, but rather to provide doctors a vacation, a tax-free social get-together, and an opportunity to play politics. He acknowledged that he was not writing papers for publication in medical journals, and he felt that in academic medical circles the failure to publish indicated failure. But T.E.B. also believed that it was not useful to add to the torrent of unimportant words flowing out to overwhelmed medical readers. When Dr. Moore offered to nominate him to the Central Society for Clinical Investigation, T.E.B. replied, courteously, that he was opposed in principle to all exclusive societies, whether they be country clubs or Alpha Omega Alpha (an honorary medical society), and that he did not belong to any exclusive society.[139]

T.E.B. had concluded that David Rogers' ideas were more in common with his own, and that at Vanderbilt he might find more common ground with his colleagues. In 1962, Rogers accepted an invitation as visiting professor at Washington University, and when he arrived in St. Louis, Brittingham felt as close to him as ever. After discussions with Rogers, T.E.B. decided that both medicine and teaching sounded more exciting at Vanderbilt. His adoration of the wisdom of his friend David Rogers was the determining factor in his move to Nashville and Vanderbilt the following year.

In a letter Dr. Carl Moore wrote to David Rogers, responding to Rogers' request for an endorsement of T.E.B.'s appointment as associate professor of medicine at Vanderbilt, it was evident that Moore understood Brittingham's feelings. He wrote that the only hesitations he had about T.E.B.'s future in academic medicine related to his research and his cynicism about most research being done in clinical departments. He added that he had argued with T.E.B. about this and believed that T.E.B. deprecated the motives of many of his colleagues.[140]

The last medical ground rounds of the academic year at Washington University in 1963 was to honor Dr. Thomas E. Brittingham. The auditorium at Barnes Hospital was packed, standing room only. The audience was aware that there had been tension between Brittingham and some of the hematology faculty. His presentation dealt with the controversy at the time as to whether Hodgkin's Disease was strictly a neoplasm. T.E.B. believed that there were immunologic, inflammatory, and even infectious components to this entity. He kept all options open, insisting that much of clinical practice could be openly questioned. The audience gave him an enthusiastic standing ovation—except for his hematology colleagues in the first row, who remained seated and clapped politely.

10

Camelot Years at Vanderbilt

"As a profession, we have done too little to demonstrate our social conscience, our commitment to our patients, and the welfare of the broader community. I believe we have a collective responsibility as a profession to be social activists."—David E. Rogers

After Tom and Dotsy's first visit to Nashville, T.E.B. wrote to his mother, "Nashville is a beautiful town in a great many ways, its residents are thoroughly satisfied with everything about it, and thus it is perhaps the most traditional, conservative, complacent, easy-going, resistant-to-change town I have ever seen."[141]

That may have been so, but things were about to change. Those changes would erupt in the 1960s and challenge the traditions of all southern cities, traditions related to race and originating before the Civil War

During the Civil War, the steamships of Commodore Cornelius Vanderbilt played a decisive role for the Union navy. But subsequently the Commodore developed a growing interest in healing the wounds of the war. Thoughts of his legacy preoccupied him. Holland N. McTyeire, a Southern Methodist bishop and a frequent guest at Vanderbilt's home in New York, informed the Commodore that, after the war, the Methodist Episcopal Church was intent on establishing a preeminent university somewhere in the South. Perhaps it was just a coincidence that McTyeire's wife and the Commodore's young second wife, Frank [sic], an unrepentant Confederate from Mobile, were cousins, but the bishop's university was the enterprise Vanderbilt decided to endow.

The Commodore explained, "It was a duty that the North owed the South, to give some substantial token of reconciliation."[142] Vanderbilt donated a million dollars, an endowment at that time greater than that of any school south of the Ohio River. He specified that the university should be located in Nashville and that the bishop should be its president. The Southern Methodists quickly accepted these terms, and changed the school's name from Central University to Vanderbilt University. It opened in 1874, the same year that Colonel Ranald Mackenzie subdued the Comanches.

Vanderbilt University came pre-equipped with a medical college, which had been operating since 1851 as the University of Nashville. The curriculum at the University of Nashville consisted only of lectures, with no laboratory or clinical exposure, as was customary at that time among the many proprietary medical schools around the country. Without the expenditure of any monies, Vanderbilt University secured a liaison with the University of Nashville and awarded its first Doctor of Medicine diploma before the university itself was formally opened.[143]

Fast forward to 1919. Vanderbilt Chancellor James H. Kirkland approached Dr. G. Canby Robinson, dean at Washington University Medical School in St. Louis and leader of that school's reorganization, to establish a state-of-the-art medical school at Vanderbilt. Robinson accepted the challenge. He was appointed Dean of Vanderbilt School of Medicine and Chairman of the Department of Medicine. He obtained funds to erect a new hospital. Then he moved to Baltimore for four years, on leave of absence from his positions at Vanderbilt. In Baltimore he observed how the Johns Hopkins Medical School and its hospital operated, and he worked with architects to design the new hospital in Nashville. He served as Acting Head of the Department of Medicine at Hopkins for a year and visited medical schools in Germany, Denmark, Holland, Scotland and England.

Robinson insisted that the hospital be located on the Vanderbilt campus. The medical school and the teaching hospital would occupy the same building. Hospital wards, research laboratories, and preclinical departments, including the anatomy dissection area and the autopsy room, would be within close walking distance of each other. This plan intended to encourage familiarity and cross-fertilization of ideas among various disciplines of the faculty. The four-story brick and limestone structure would consist of three corridors perpendicular to two corridors, enclosing two courtyards. If seen from above, it resembled a two-level tic-tac-toe pattern.

Robinson's plan for Vanderbilt included only a few full-time faculty members, supplemented by a highly competent part-time volunteer clinical faculty. The part-time faculty would do the lion's share of the clinical teaching, but their livelihood would depend on private practice. This concept reflected what Osler had instituted at Hopkins.[144]

In 1928, Robinson left Vanderbilt to reorganize and build the New York Hospital-Cornell University Medical College (twenty-two years before Tom Brittingham would arrive there). But with medical stalwarts such as Sidney Burwell[145], Tinsley Harrison[146], Rudolph Kampmeier,[147] and Hugh Morgan[148] on board, the Vanderbilt Department of Medicine flourished.

The early years of success came to an abrupt halt when the country entered into World War II. Seventy-five percent of Vanderbilt's full-time medical faculty left to serve in the military. Department of Medicine Chairman Hugh Morgan went on leave of absence to serve as Chief Consultant in Medicine to the Army's Surgeon General. The Department limped along, with the loyal support of the part-time faculty. These unpaid volunteers covered teaching on the wards and clinics, conducted some of the courses and conferences, and gave lectures.[149] Dr. John Youmans paid tribute to them, "In my opinion, in the years from 1927 to World War II, Vanderbilt Medical School possessed the best volunteer faculty of any medical school in the country."[150]

After the war, a new era was beginning. Scientific and technological advances resulted in increasing specialization and insidiously eroded the central role of bedside medicine. Technological progress provided valuable objective and concrete data and outshone the more subjective results achieved by the physician's history, physical examination, and clinical judgment. General internal medicine, a specialty limited to adults but overlapping with primary care, was gradually diminishing as a vocation. By the late 1950s the majority of Vanderbilt students chose to become specialists.

This progress also stimulated growth, which changed the atmosphere of the school and hospital. Affiliation with the Nashville Veterans Hospital and the Nashville General Hospital expanded teaching opportunities but diluted the cohesion of faculty and students, who no longer worked under one roof. Dean John Chapman characterized expansion, "We're getting so big and so scattered that every pigeon has a hole."[151]

When Morgan, a heavy smoker, developed emphysema, he delegated the duties of running the Department of Medicine to a triumvirate of his colleagues. He appointed Dr. Rudolph Kampmeier to handle the administrative work, Dr. Elliot Newman[152] to be in charge of the third-year medical students and the house staff program, and Dr. Grant W. Liddle, knowledgeable about grants and the National Institutes of Health (NIH), to supervise research efforts.[153] These departmental leaders served as the search committee to identify candidates for Morgan's successor. It was no easy task. Vanderbilt had a glowing reputation in teaching and research, but few immediate resources. The endowments of the 1920s had diminished, the limited faculty faced hurdles to compete effectively for grants, and several areas of the physical plant needed refurbishing. A young, eager, infectious disease specialist at Cornell Medical College attracted the attention of the search committee. They recruited him not only as an agent of change to refashion the department of medicine, but almost as a medical missionary for the entire school. This young man—who seemed ahead of his time—was Dr. David Rogers, T.E.B.'s best friend. Rogers was the man to shake up the status quo.[154]

Rogers assumed the job as Chairman of the Department of Medicine at Vanderbilt in 1959. At age thirty-three, he was the youngest in the United States to hold such a position. Some called him the "Boy Wonder," as he set out to recruit a young faculty who would transform Vanderbilt from a regional center into one that competed for research grants, faculty, and house staff with national leaders.

T.E.B.

Photo courtesy of Special Collections,
Vanderbilt Eskind Biomedical Library

Dr. David E Rogers

David Rogers had a profound sense of right and wrong. He was passionate and uncompromising in advocacy of his principles, no matter how unpopular his position. Shortly after his arrival, Rogers' strong views embroiled him in controversy, the first of which he initiated. At Cornell, only the house staff could write orders in the chart for the patients under their care, even when the patient had a private physician. It was assumed that the private physician and the house staff would have close contact and work as a team, and each decision about a patient would be a learning activity. A few months after his arrival, Rogers initiated a similar system at Vanderbilt. It was like an unexpected executive order. It backfired.

The part-time faculty, who had played an indispensable role during the war years and still provided most of the clinical teaching in the wards and clinics, were offended, and most rebelled. They felt strongly that they should be able to write orders on the private patients they had admitted.

Rogers' directive also required the part-time physicians to spend more time in the hospital and less time in their offices, where they earned their livings. Many stopped sending their patients to Vanderbilt and began to admit them to nearby community hospitals. The virtual disappearance of the part-time faculty changed the culture of Vanderbilt Hospital and damaged town-gown relations for decades. The full-time faculty had to assume responsibility for the care of all hospitalized patients—and bill them. Many felt this compromised their research activities.

The second major controversy stemmed from Rogers' dedication to social action. He joined this controversy voluntarily. In 1960, the Lawson affair, as it was called, engulfed not only the Vanderbilt campus but the entire city of Nashville. Social justice was the passion of Reverend James Lawson, a thirty-year-old African-American Methodist minister and a graduate student at Oberlin School of Theology. He had gone to jail as a conscientious objector during the Korean War. Then, as a missionary, he had studied philosophies of non-violence in India. When he returned to Ohio, he met Martin Luther King Jr., who urged him to come South and join the civil rights struggle. Lawson transferred to the Vanderbilt Divinity School.

He trained young black protestors in non-violent Gandhian strategies of civil disobedience. He instructed them to accept verbal and physical abuse—and even arrest—without fighting back. He organized sit-ins at downtown lunch counters that refused to serve blacks. Scores of black students (and a few white students) took part in the sit-ins, while angry whites harassed them. Local newspapers portrayed Lawson as an outspoken leader of the sit-ins. In the midst of nationwide media publicity in March, 1960, the conservative-minded Vanderbilt Board of Trust and the Vanderbilt Chancellor, Harvey Branscomb, expelled Lawson from the Divinity School for his role in the movement. Lawson's expulsion sparked soul-searching about Vanderbilt University's mission. Was it a major center of learning or, as critics put it, "a Southern finishing school?"[155]

More than half the Divinity School faculty members submitted their resignations. Several other faculty members and a half dozen medical school professors, the most influential of whom was David Rogers, also submitted resignations. The resignations from the medical school meant the possible loss of millions of dollars in research funds. The latter was a cause for concern to the Board of Trust.

At this point, Harold Vanderbilt, the great-grandson of the Commodore, well into his seventies, took an active interest. He had been on the board for many years, but had never involved himself emotionally with the southern university that bore his family name. The adverse publicity had become a family embarrassment. Harold Vanderbilt proposed a solution, which the chancellor and Board of Trust eventually accepted. Lawson was reinstated—although he chose to continue his postgraduate studies at Boston University. The faculty withdrew their resignations. The strong protests of Rogers and other faculty members had been effectual in turning the tide. Subsequently, Vanderbilt Hospital was entirely desegregated, largely as a result of Rogers' role on the executive faculty of the medical school.

Also high on Rogers' list of plans was an effort to boost Vanderbilt's role at the Nashville Veterans Administration Hospital, which planned to move to a location adjacent to Vanderbilt Medical Center. Rogers needed someone who would develop the VA department of medicine in an academic direction. Dr. Roger Des Prez, an assistant professor at Cornell, had a broad view of medicine, particularly at the overlap between pulmonary and infectious diseases. With experience gained at an Indian reservation, he had become an expert on tuberculosis. Dr. James Snell, a Vanderbilt pulmonologist then studying at Cornell, knew that Des Prez had reached a crossroads in his career and was considering going into private practice to support his large family. Snell notified Rogers of Des Prez' possible availability, and Rogers recruited this talented physician-scholar ro Vanderbilt.[156]

Photo by Dr. Eric Dyer

Dr. Roger Des Prez

Regarding the eventual success in recruiting T.E.B., Rogers longed for an individual with superior teaching and clinical skills who could serve as his alter ego and share the load of running the department. He felt that attracting Tom Brittingham was "like filling an inside straight in a poker hand." Rogers could focus on building the department and addressing wide-reaching issues in medicine, while T.E.B. would devote himself entirely to teaching and patient responsibilities.

Rogers created T.E.B.'s dream job. His title was Vice-Chairman of the Department of Medicine, but his administrative responsibilities were limited to when Rogers was out of town. His office was right across the corridor from Rogers' office and the departmental conference room. T.E.B. had no requirement to apply for research grants, unless he wished to. He interviewed all applicants for internship. His opinion usually determined whether or not the candidate was invited to come to Vanderbilt. In other words, he virtually handpicked the medical house staff. He was in charge of the clinical teaching of third-year medical students, of which the most important part was their six-week clerkship on the medical service.

In essence, T.E.B. assumed dominion over medical teaching at Vanderbilt. He was under no obligation to create revenue by charging private patients. He did, however, see multitudes of such patients, including faculty members, house officers, medical students, and their wives and families. He also held a weekly clinic at the Nashville General Hospital, where his practice consisted of indigent patients.

The early 1960s at Vanderbilt were invigorating years. The arrival of stimulating physicians such as Roger Des Prez, Tom Brittingham, Glenn Koenig in infectious diseases, and David Law in gastroenterology produced an atmosphere of enthusiasm and collegiality in the Department of Medicine. Every Friday afternoon Rogers invited the medical faculty for talk and drinks in his office. Attendance was expected. It was another custom he brought from Cornell.

It was possible at that time for the entire faculty of the Department of Medicine to fit into someone's living room, and often it was Tom and Dotsy's. David Rogers' youth, exceptional clinical skills, and personal warmth created an optimistic and congenial faculty and house staff, which welcomed T.E.B.'s unique gifts. Some referred to these times as the "Camelot Years" at Vanderbi

*****

T.E.B. bounded into his new job with typical enthusiasm. His first teaching assignment was on the C-3100 ward (now part of Medical Center North). Charley O'Donovan was the intern and Bill Schaffner was the resident. In the summer of 1963, Vanderbilt Hospital was not air-conditioned, and Schaffner remembers sweat dripping off his nose onto the patient notes he wrote in the tiny doctors' workroom adjacent to the nursing station. When Dr. Brittingham and his family moved into their newly acquired home in Nashville, they installed central air conditioning and donated the old window air conditioners to Vanderbilt Hospital. Several were placed in the doctors' workrooms. T.E.B. had scored with the house staff before even meeting them. But more important, Dr. Brittingham's reputation—of pursuing every detail about a patient—had preceded his arrival.

O'Donovan and Schaffner diligently prepared their patient presentation to their new mentor. When O'Donovan presented his patient's history, he mentioned a previous thyroid problem, which he and Schaffner had not thought relevant to the current hospitalization. T.E.B. interrupted and asked about the details of the thyroid problem. They told T.E.B. what they knew, but Dr. Brittingham had much more information. O'Donovan and Schaffner were unaware that T.E.B. had met with the patient and her family the day before. He had taken his own medical history, and had examined her thoroughly. When he learned of a previous hospitalization at Baptist Hospital for a thyroid problem, he went to the medical records department there and reviewed the patient's chart. The intern and resident had been scooped—not by a new research discovery, not by a scholarly article in the medical literature—but by T.E.B.'s tenacity in getting all the information about their patient, which was their responsibility.

"I remember vividly that challenging first meeting with T.E.B. It was over fifty years ago," recalled Schaffner. "The *joie de vivre* he exhibited in the practice of medicine was an energetic, infectious force. It drew you in; it called upon the best in you; it provided an example to emulate. You wanted to engage with him; you didn't want to disappoint him; you never wanted to be less dedicated than he."[157]

T.E.B. never used expressions such as *patient-centered* or *holistic care*, but his actions showed what they meant. He taught by example. He never expected anything of the house officers that he was not willing to do himself. The young doctors' greatest fear was that Brittingham would workup one of their patients. Inevitably he would discover important information that had been missed. After T.E.B. was seen leaving a patient's room one evening, the student rushed to the patient's bedside to ask, "What did you tell that doctor who just left that you didn't tell me?"

Charles Mayes, a resident at the time, said, "If I had time, I would stand by the curtain drawn around the patient's bed and listen to T.E.B. interview him. Everything the patient told him, T.E.B. just lived it with him, he sounded like he felt the same way the patient did."[158]

Brittingham dominated clinical teaching at Vanderbilt for the next 17 years. During those years his principles of patient care and medical education matured. He applied a sense of deep interest, genuine concern, and personal responsibility to every patient he encountered. He charged the house staff to be responsible for every aspect of a patient's care—health, comfort, emotional security, and social situation.

Agnes Fogo presented a case one Saturday morning at the medical students' rounds. The patient was weak and couldn't walk. T.E.B. knew much more than Agnes about how weak the patient was. He had drilled down into the man's everyday life and had found exactly what he could and could not do in his normal activities. Animated, he showed Fogo how she could have gotten much more specific information. The students learned that in order to know what was different when the patient was ill, you must first understand what he did when he was well. And then T.E.B. said, "Well, I'm just a simple country doctor."[159]

T.E.B.'s method of teaching, whether targeted at students, house staff, or practicing physicians, did not depend on the transmission of facts. It was not designed to impart an organized body of information on a defined subject. He never presented a table or a graph. He did not use slides. He never concentrated on the latest diagnostic tool or the most recent therapeutic wonder. He did not speak about populations, averages, or statistical significance. In short, his teaching method was radically different from typical medical education.

Instead, Brittingham centered on the clinical details of a single patient. At the bedside, at morning report, at professor's rounds, at death conferences, at clinical pathologic conferences (CPC's), or at Medical Grand Rounds, the focus was always the careful examination of one case at a time. Either a patient would be brought into the conference room, or T.E.B. would lead a group to the patient's bedside. He delighted in hearing patients describe their symptoms in their own language. He stressed that the patients were the most important teachers of medicine.

T.E.B.'s teaching skills often provided great theater. Dr. Des Prez asked him to lead a conference at the VA Hospital about ulcerative colitis. The session began with T.E.B. outlining the basis for making a diagnosis of ulcerative colitis. He frequently glanced at a stack of journal articles he had piled by the podium, occasionally quoting passages from some of them. While he looked for this or that article, he imparted that the authors must be very smart, since the diagnosis seemed vague and confusing for a simple man like himself. Then he presented five cases, each having carried the diagnosis of ulcerative colitis for many months or years. Two ultimately turned out to have colitis due to histoplasmosis, two had amoebic dysentery, and one had tuberculosis.

He did this sort of conference often, chuckling about how smart these authors were and how hard it was for poor Dr. Brittingham. Each quote from the literature at first seemed sensible, but gradually defects in the argument peeked out. His forte was the skillful destruction of simplistic ideas and presumptions. One came away from such a session with the conviction that protection of the patient required all the thoroughness and thoughtfulness that one could muster.[160]

The students and house staff watched T.E.B. talk to patients and family members. He understood that people in the hospital were emotionally distraught. Instead of approaching a new patient and asking about his belly pain, T.E.B. asked personal questions before he inquired about the clinical history: "Where are you from? Tell me about your job, your family. Why do you think you're here? What do you think is wrong with you?" He said, 'The patient can tell you what's wrong, if you just listen to him.'"[161]

Before the third-year medicine clerkship, T.E.B. not only knew the students' names but also an uncanny amount of detail about their personal lives. He studied student files at home the evening before he met them. Dr. Marvin Gregory stated that Brittingham "knew your name, your wife's name, and your dog's name" on the first day.[162] He knew not only their professional lives, but their personal lives—their health, their background, their family, their marriage. He possessed an intuitive understanding of people and was able to judge them like his Uncle Watt judged cattle. Terry Moberly, the longtime foreman at Lambshead Ranch, said of Watt, "He could look a cow in the butt and read her mind."[163] T.E.B. could do it just as well with medical students. Dr. Eric Dyer said, "When I was in his presence, I had the sense that he understood my personal and professional being better than I understood myself."

On the first day of the medical clerkship T.E.B. distributed to each student a detailed guide of what he expected of them during the next six weeks:

MEDICAL CLERKSHIP

> I consider that the first and greatest responsibility of your Medicine clerkship is to obtain a complete and accurate history, perform a complete and accurate physical examination, and to record fully the results of both. To obtain a complete history, it is wise to use multiple sources, to read every word in all volumes of the patient's old Vanderbilt chart (summarizing the pertinent information therein), and often times to read the patient's record at other nearby hospitals. One becomes a good physician (whatever specialty one chooses) by first learning how to make detailed and correct observations, then by learning how to make a diagnosis from the observations ... and then finally by learning how to make the correctly diagnosed illness get better. If you learn how to obtain a flawless history and physical examination during your third year in medical school, I think you will have done well. To achieve this is

easily within the capabilities of each one of you. No one of you can tell me right now that it is a waste of time to be careful in medicine. You will first have to try making careful detailed observations, and only then can you decide whether or not this is worthwhile for you. I shall be entirely satisfied to have you make this decision for yourselves AFTER you have worked up in detail the ... patients of your Medicine clerkship. Strange as it may seem, you probably have more time now than you will ever have again, once you have graduated from medical school. It is only by being thorough now that you can learn how to be selectively thorough later, i.e., where you can safely take shortcuts. What you learn in becoming a good medical observer will be useful to you always. Contrast this with the "facts" which you learn from your books. It is said that half of what we are taught in medical school will have been shown to be wrong during the first ten years after we have graduated.

I expect that all peripheral blood film examinations and all urinary sediment examinations on your patients, old or new, will be performed by you and initialed accordingly in the chart. Examination of the peripheral blood film and of the urinary sediment are powerful diagnostic tools, much as is physical examination of the heart. The power of these tests is unappreciated by many physicians, principally because the latter are incompetent in examination of the peripheral blood film and urinary sediment, think of them as "scut" to be relegated to a laboratory technician. The blood film or the urine may contain the critical data for a given clinical problem. The good doctor, whenever possible, examines the critical data personally rather than leaving it for someone who is less educated, is paid less, and is less motivated. You can only become competent at blood film and urine sediment examinations by doing a large number of them. You will probably no longer have time to teach yourselves these skills after you have graduated from medical school.

A section called Formulation should be incorporated into each history you write. It entails the preparation of a well-organized, concise, and logical discussion based on a thorough knowledge of the patient's symptoms and signs and of the diseases which they signify. It should indicate why you are choosing one diagnosis as most likely rather than another (e.g. in a patient with chest pain, why do you think he has a pulmonary embolus rather than a myocardial infarction or aortic dissection? in a patient with abdominal pain, why do you think he has peptic ulcer disease rather than cholelithiasis or pancreatitis?). It should also make note of the grossly atypical clinical features which suggest that your diagnosis may be wrong, and of how you are attempting to explain these features. In writing this section, one is trying to teach oneself to think and to write clearly, also to be ever aware of the possibility of ERROR in the formulation of the patient's problem. Greater skill is demanded in writing this section than in any other portion of the record. Do <u>NOT</u> under ANY circumstances write a formulation longer than one page, a longer one will simply waste your time and make others less likely to read what you have written. Usually your formulation should be limited to one side of one page, and often it can be shorter than this.

On first coming to the Medical Service, you will be assigned three patients already in the hospital. In addition, you are to work up at least three new medical admissions each week, i.e., you will have written up 18 new medical patients during your six-week stay here. Your assistant resident is responsible for going over each workup with you. Each of your new admissions must be completely written up and your signed note be in the patient's chart within 24 hours of his admission ... Please keep a list of all patients worked up while here and leave a copy with me at the time you leave the Medicine service. Include history number with the patient's name; this will enable me to go to the record room and look at some of your notes if I wish, and it will also greatly facilitate your own long-term follow-up of your patients.

It is highly desirable for you to accompany your patient to major procedures, e.g., surgical operations. Record regular progress notes on all of your patients, both old and new. I expect you to spend time with all of your patients after the initial workup, establishing rapport with them, obtaining an extensive personal history by indirect and easy means, and in showing the patients that you have a keen interest in them as people and as friends as well as diseases.

On Saturday mornings at 9:00, two patients will be presented by the students to me ... You need not present an "interesting" patient on Saturdays, may feel free to present patients with simple senility or anxiety or whatever you wish. Your attendance at this is strickly (sic) optional, for I believe your learning occurs by what you do, not by what I say to you. Never take longer than 10 minutes to present a case, no matter how long and complicated the case seems ... Since you are presumed to know the case better than does anyone else, you are best suited to determine what the central facts are and to distill these for your listeners. Ultimately, you should be able to present any case in five minutes or less.

Should one of your patients die, you are expected to leave your regular ward duties or your home and attend the complete autopsy, just as if you were performing the latter yourself.

Check carefully with each of your fellow students for abnormal physical findings they have seen on their services. By confirming all of these findings for yourselves, you can make yourselves into adequate physical diagnosticians.

Dr. Carl Moore, the late Chairman of the Department of Medicine at Washington University, told me that he found it necessary to read medicine three to four hours every day, 365 1/4 days yearly. If he found it so necessary to read in order to remain competent, we probably need to read too in order to become competent. Half of your evenings are unscheduled so that you may have the opportunity to read about the patients you have seen. Read and think extensively about their disease problems; you are expected to be able to contribute many useful diagnostic and therapeutic suggestions toward the total care of your patients.

You will find that your time on the Medicine Service is relatively unsupervised, i.e., you will be principally responsible for your own performance. This seems appropriate, since I feel confident that each one of you has more understanding of wherein lies his or her best interest than does any outsider, particularly me. It also seems appropriate because your future education will be almost all on your own, unsupervised. It is helpful to be able to evaluate one's own performance. If you are doing a good job, you should find that you miss nothing of any importance in the history or physical examination of any of your patients, i.e., your intern, resident attending physician consultants, etc. find nothing that you did not already know ... You should also find that you are able to add significant information (not elicited by your intern, resident, etc.) to the workup of each one of your patients. If you are not able to do these things, you need to do better ...

You will be considered by me to be personally responsible for the welfare of your patients. You are expected to help inculcate in your patients the feeling that the patient is always right, that nothing is too good for the patient, and that you consider it a privilege to be able to care for him.

T.E.B.

Reread these pages occasionally to remind yourselves of what you should be achieving. If you are able to do all of the above, you will probably find that your clerkship has been well spent and a satisfying experience. You will have been a great help and inspiration to us, made us better doctors for your presence. You will also probably have learned that for each one of us the primary teaching responsibility now and for the rest of his life lies within himself or herself.

T.E.B.[164]

*****

A typical Saturday morning session with the medical students follows. After Lewis Schrager, a third-year medical student, presented the case, T.E.B. said, "Lew, I've spent some time with Mr. K. myself. Would you mind if I told the group what I've learned?"

Schrager nodded his approval, and T.E.B. continued, "Now, as Lew said, Mr. K. entered the hospital for an aorto-femoral graft, and is now convalescing from the operation. He's had a hell of a year, wouldn't you say, Lew?"

"Yes, sir," Lew agreed.

"And he's had problems that go back further. In fact, he told me that he began feeling unwell around two years ago. Not altogether sick, mind you, just not well. He felt like he had *the blahs*, you know, couldn't concentrate at work, felt kind of listless." After a pause, T.E.B. continued, "His friends said that he was probably depressed, so he went to see his doctor, who said maybe Mr. K. was suffering from a mid-life crisis. This made Mr. K. feel worse, because he'd never had any *head problems* before. Well, things got worse. At his wife's urging, Mr.K. went back to his doctor. This time on physical exam the doctor found a hard nodule in the prostate. A biopsy of the prostate showed prostate cancer, with metastases. Orchiectomy, the standard of care then, was performed. Surgical resection of the testes decreases testosterone production, which stimulates growth of prostate cancer."

T.E.B. let that sink in, then resumed. "Over the next few months Mr.

K. became more depressed. He said everyone now told him that feeling depressed was perfectly understandable, how else should someone feel after a diagnosis of metastatic cancer and having your balls cut off."

The students stared at the floor. They couldn't risk T.E.B. seeing their suppressed smiles. T.E.B. went on, "Well, Lew, you're a pretty compulsive fellow and I'm sure you know the rules here. Review for us the findings of all the pathology reports, from the very beginning."

In fact, Lew and all the students did know the rules—that every piece of resected tissue had to be reviewed personally with a pathologist before presenting a case to T.E.B. And with a compulsion born of fear, Lew had indeed reviewed all of the tissue removed from Mr. K., going back to his tonsillectomy at age four. Conspicuously absent, however, was any report in the chart of Mr. K.'s testicles. Since there was no path report, Lew assumed there was no issue—after all, they were just normal testicles that had only been removed to decrease the growth of the cancer.

Lew started, 'The initial path report on the tonsils ...''

T.E.B. interrupted. "That's fine, Lew. Now tell us about his testicles."

Lew started sweating. "Well, uh..."

T.E.B. said, "You know the rules, Lew. You did review the slides of Mr. K's testicles, didn't you?"

Lew said, "Well, no sir, I didn't."

T.E.B.: "And why is that?" T.E.B. asked.

Lew, "Well, sir, because I couldn't find a path report in the chart, so I didn't think it had been done."

T.E.B. said, "Lew, if you had YOUR balls cut off, wouldn't you want them examined by a pathologist?"

Lew, amid suppressed chortling of the surrounding students, admitted, "Well, yes sir, I guess I would." T.E.B., chuckled, "I'll bet you would. I would, too. Lew, I'm glad you're an honest man, because I didn't find a pathology report about his testicles either. Not until last night, anyway. I had a hell of a time convincing one of the security guards to let me into the pathology offices at 1:00 am on a Saturday morning, but I got in, and you know what I found? The pathology report, buried in one of their files. The duplicate never made it to the chart." T.E.B. triumphantly waved the characteristic yellow report high over his head.

"So, Lew, would you be interested to see what it says?" T.E.B. chuckled again, "It says, normal testicles. How about that? But, Lew, you know who signed this report? The chairman of the pathology department, that's who signed it. Lew, you know what that makes me think?"

Lew answered, "No sir, what's that?"

"It makes me think that a pathologist never looked at those testicles! Now, if it were a pathology resident who signed off on it, I'd know that every inch of those slides had been carefully scrutinized, but the chairman isn't going to spend any time examining incidentally-obtained tissue. You've got to question authority, Lew, something I learned a long time ago."

Then T.E.B. took all the students to a multi-head microscope and demonstrated a small focus of obvious granuloma (a chronic infection) in the testis, which he had discovered at 3:00 a.m. that morning. When stained for fungi it was found to be teeming with histoplasmosis organisms, a fungus commonly found in the soil of the Ohio and Mississippi River valleys. It frequently causes infections in the lungs. If it becomes disseminated throughout the body, a serious, sometimes fatal illness results. Disseminated histoplasmosis may be hard to diagnose without biopsies of several organs.

T.E.B. concluded by stating that Mr. K.'s so-called *depression,* which had indirectly led to the diagnosis of his cancer and to his orchiectomy, was really caused by disseminated histoplasmosis. In other words, the incidental finding of the prostate cancer was not the cause of the patient's symptoms, nor was depression—the patient had been suffering from an undiagnosed generalized fungal infection.

Dr. Lewis Schrager, who was on the staff of the National Institutes of Health (NIH) in 1993, wrote, "I believe there are more than a few lessons buried within this single episode of my life, which occurred over a period lasting less than an hour, many years ago. T.E.B., the teacher of these lessons, was an extraordinary individual whose memory could not be better preserved than by dedicating a learning center in the hospital in his name."[165]

On a December night, faculty members covered all the hospital units and emergency rooms, allowing the interns and residents to have a Christmas party. T.E.B. volunteered to cover the emergency room at Nashville General Hospital. That night T.E.B. demonstrated what being a doctor was all about. He showed up wearing his usual Oxford shirt, striped tie, Brooks Brothers charcoal gray slacks, and cordovan shoes, and the intern went off to the party. When the intern came to relieve him the next morning, T.E.B. was sleepless, and his white shirt was splattered with blood and vomit.

After T.E.B. left, the intern breathed a sigh of relief. He was still tired from the party, and the emergency room did not appear very busy. A few hours later, a patient showed up, holding one of T.E.B.'s barely legible handwritten notes. "His left lung sounds worse, he needs a chest X-ray," said the note. Over the next few hours several other patients trickled into the emergency room, each bearing a note. One note said, "His pain is no better, he probably needs to be in the hospital." And so it went throughout the day, as T.E.B. made house calls in the nearby housing projects in his rusty Thunderbird, following up on the patients he had seen the night before.[166]

Dr. Rick Davidson told me, "T.E.B. taught us to practice in an *idealized, non-realistic way*. These lessons were important for learning, even if difficult to replicate."[167]

"I fell in love with medicine early, when I was a medical student at Hopkins," said Dr. Bill Stone. "I fell in love with people who were intensive about it and cared. T.E.B. cared."[168]

Some of Brittingham's most memorable teaching moments occurred away from the bedside or the conference room. Steve Nace, a third-year medical student, was eating lunch alone at a table for two in the hospital cafeteria. He was surprised to see T.E.B. coming his way and take the opposite seat. A professor eating lunch with a medical student! Brittingham recognized that Nace was a little uncomfortable with the situation, so he made small talk for a few minutes. Nace mentioned something about how hard it was to be a good doctor. Brittingham immediately changed focus, looked Nace in the eye and said, "Steve, there are three things you've got to get comfortable with. The first is that in spite of everything you do, you will never know what's wrong with some of your patients. The second is that in spite of everything you do, some of your patients will get worse. The third is that some of your patients will get worse because of what you do."

Nace said that this was the most important lesson he ever learned at Vanderbilt.[169]

Dr. Clif Cleveland told me, "T.E.B. had an emphasis on narrative—he knew how to elicit and interpret it. He was at his absolute best pulling people into dialogue."[170]

Dr. Jim Loyd stated, "T.E.B. taught the students to gather and review primary data personally, not to make assumptions. He said to pay attention to what you found yourself, to how the patient looked to you Only personally obtained data could lead to making a confident diagnosis."[171]

Dr. John Sergent characterized Brittingham as the ultimate iconoclast, challenging every old convention and every new test. He said that much of the *information* T.E.B taught was *wrong*. Some of the examples cited included Brittingham's not using chemotherapy for lymphoma and not using steroids for giant cell arteritis. He added that T.E.B. thought Hodgkin's disease was an infection, that diabetic kidney disease was due to an immune reaction to insulin, and that then-new coronary artery catheterization was unsafe. He was proven wrong on each of these issues. "But," Sergent concluded, "Brittingham taught us something that you cannot get from books or journals. He taught us that we were completely responsible for our patients."[172]

11

Master of Medicine at Vanderbilt

"Education is not the filling of a pail, but the lighting of a fire."
—William Butler Yeats

For the next 17 years, a chinning bar in T.E.B.'s garage at the entrance to the kitchen would be his principal—virtually his only— exercise. Dotsy had given him the chinning bar and a new Schwinn bicycle for Christmas in 1963, a month after President John F. Kennedy was assassinated in Dallas. He could jump up and grab the bar and chin himself twenty times or more. He rode the bicycle with his children on weekends, which were also packed with social events.

Most of Tom and Dotsy's original friends in Nashville were medical colleagues and their wives. In addition to David and Corky Rogers, they socialized with Dr. Tom Paine, chief of the medical service at Nashville General Hospital, and his wife Grace, whom T.E.B. called "saintly people." The Brittinghams also became close friends with Vanderbilt Chancellor Alexander Heard and his wife, Jean. They renewed old friendships with Dr. Guv Pennington and his wife Phyllis. Guv had interned at New York Hospital a year after T.E.B. and was now in practice. His active participation in teaching at Vanderbilt prompted T.E.B. to surmise, perhaps prematurely, that *all is serene with the private practitioners with no more bitter warfare between town and gown.* But when T.E.B. and Dotsy attended a dinner at the home of a local neurologist, T.E.B. thought the neurologist emphasized the gulf between private practitioners and academic medicine. The neurologist remarked that the man in practice felt that he was working hard to earn money and then pay big taxes, which were used to support the parasites in academic medicine.

T.E.B.'s letters to his mother often contained ambivalent thoughts about his career and the personal challenges of his idealism. He had the uncomfortable feeling of not doing what he ought to be doing. This would be a recurrent theme for the rest of his life. He suffered pangs of guilt that he had spent so much money by moving into a large house, when so many people had so little.

David Rogers was aware of T.E.B.'s qualms and tried to reassure his friend. He said that having T.E.B. join the faculty had been the most satisfying milestone in his four years at Vanderbilt. He wrote to T.E.B., "I would like to see you satisfied with what you are and what you do and have fun with it as well."[173]

Soon T.E.B. was having fun and was so busy he feared he was neglecting his family. He wrote to his mother, "Dotsy has convinced me that children need a father as well as a mother; I had always thought the father somewhat superfluous." One day he came home early and played with the children in the yard; he thought it worked so well that he should do it much more often. David Rogers and T.E.B. talked at lunch about how much less they gave to their children than their parents had given to them. "Am aiming to do better," T.E.B. wrote.[174]

Though he was sympathetic to justice in politics, Brittingham was not a social activist like David Rogers. He did not participate actively in the election process of 1964. In the fall of that year T.E.B. and Rogers took eight children to a football game at Montgomery Bell Academy. T.E.B. described the atmosphere as similar to an Ivy League game. He characterized the MBA people as uniformly Goldwater supporters. When one of his children announced at school that Dotsy and he would move to Russia if Goldwater was elected, he was horrified. "You can imagine what our Nashville reputation is like now," he wrote.[175]

THE INTERN SELECTION PROCESS

T.E.B. handpicked the Vanderbilt medical house staff. In applicant interviews, he asked questions about the applicant's family, where he grew up, what his goals in life were. He wanted to know each of them as a person. He had a good eye for judging the things that were not in a transcript.

When Karl Vandevender completed his first year as a medical student at the University of Mississippi, he took a summer job at a little clinic in Sewanee, Tennessee, about ninety miles southeast of Nashville. One day a message arrived at the clinic about a woman who was sick at nearby Jumpoff, Tennessee. One of the clinic doctors asked Karl to check on this lady. Welcoming any chance to be of help, Karl drove to Jumpoff, where he found an elderly woman who appeared severely ill. "That's all I knew, that she was sick," said Karl, "So I picked up the phone and called the emergency room at Vanderbilt."

It so happened that Dr. Jim Sullivan was the resident on duty in the emergency room, and Karl knew Sullivan from college. "Sullivan told me to send her down in an ambulance and Vanderbilt would take care of her. And that was the last I heard of her," Karl said.

Three years later Karl interviewed at Vanderbilt for an internship in internal medicine. He waited in Dr. Brittingham's modest office. Upon entering, T.E.B. said, "You're Vandevender, I've met you somewhere before."

Karl replied, "I don't think you have. Without doubt I would have remembered that."

T.E.B. began shuffling through stacks of little index cards, then chuckled, "I knew I knew you. It says right here that you made a house call in Jumpoff, Tennessee, and referred the patient to Vanderbilt. You sound like my kind of guy. I think you'll fit in here fine."

"T.E.B. offered me an internship position on the spot, which I accepted." Karl said. "But once at Vanderbilt, I never felt I knew him as a person. I knew him as a teacher and mentor, but I could not go up to him and say, 'Hey Tom, would you like to go out for dinner?' I think there was an appropriate formality there, he was not setting out to be our buddy, he was setting out to be our teacher."[176]

In 1973, when discussion regarding a change in the on-call schedule from every other night to every third night was intense, T.E.B. wrote, "We got some very good talent for medical interns this coming year. It was gratifying to me because many here argued we would need to make things easier in order to get interns. I think it might be just the opposite. The way to attract superb talent is to have a program hard enough to stretch them—physically, psychologically, intellectually."[177]

T.E.B.

STRUGGLING STUDENTS AND HOUSE OFFICERS

No one was more committed to helping the house staff succeed than T.E.B.. He had chosen each of them, and he was committed to their success. Likewise, the Vanderbilt medical students were selected with care, and it was rare for any of them to fail for academic reasons. When a student's performance threatened his continuation at Vanderbilt, the failing student was a topic at a faculty meeting. T.E.B. was forceful in advocating for the failing student. He would ask fellow faculty, "Have we given him every opportunity?" When the answer came back "Yes," T.E.B. might ask, "Could we give him another opportunity?"[178]

While a medical student, Fred Callahan had an intense interest in sailing. He wanted to join the Harbor Island Yacht Club on Old Hickory Lake. His application required recommendations from current club members. Callahan discovered that three members of the Vanderbilt Medical School faculty, neurosurgeon Culley Cobb, pediatrician Jan Van Eys, and internist Tom Brittingham, were members of the yacht club. T.E.B. was not a sailor but had joined the club at Dotsy's insistence. As a second year student, exposed so far to only pre-clinical faculty, Callahan knew none of these clinicians.

Callahan approached Brittingham for an endorsement. "Within minutes It was apparent I had made a capital error," Callahan said. "He gave me his philosophy of medicine, which left no time for idle afternoons leaning into the wind on Old Hickory Lake." As T.E.B. handed the signed application back with his characteristic chuckle, he remarked, "I hope, Dr. Callahan, that you have fun sailing your yacht while the rest of us country boys are over here taking care of your patients."[179]

When Dan Canale was a fourth year medical student, he was planning to go into medicine or pathology. While working at Nashville General Hospital one night, he took a patient's history, as much as he could get from a patient who was lethargic and confused. Then he tried to examine the patient, but the patient was uncooperative and pushed him away. After a lengthy struggle, he still had no idea what was wrong with the patient. Next, he did the urinalysis and the blood count and looked through his microscope at the blood cells he had smeared and stained on a glass slide. After spending only fifteen minutes in the laboratory, he knew the diagnosis. He realized *I can do this part better. I can still do good, but I don't have to dedicate 100% of my time to medicine.*

T.E.B. told Canale, chuckling, "Danny, internal medicine isn't like accounting, you can't go home with the numbers all balanced each day. You've got to be willing to run some debits."

Dr. Dan Canale didn't choose internal medicine as a career, because he thought he would have to practice like Brittingham. He became a distinguished pathologist. "I could never measure up. T.E.B. set the bar so high, I felt it would be my whole life," he said.[180]

In July, 1964, T.E.B. ran the medical service for six weeks while David Rogers was on vacation. During that period one of the interns made a colossal error on a patient in the emergency room. T.E.B. was ambivalent about handling it. He agonized, "The mistake has to be pointed out in such a way that the boy who made it will not be destroyed and that others can learn from it. I don't know if I will be able to do this."[181]

Brittingham was intolerant of any physician, especially himself, who put his needs before those of his patients or colleagues. A resident took a short leave of absence to visit his mother, ill with cancer. The resident did not return on the date agreed, nor did he call. T.E.B. found this action intolerable.

One resident drew attention to himself by having his intern push him around in a wheelchair during morning rounds. One morning in the cafeteria at the VA Hospital he bragged that he had not examined a single patient in the last two months. Dr. Roger Des Prez, chief of medicine at the VA, overheard this braggadocio. The resident had a brief conference with T.E.B., then quietly left Nashville in mid-year.[182]

When T.E.B. observed or learned about deficiencies in the performance of a house officer, he wrote a letter to the intern or resident involved. With firmness but in a respectful manner, he pointed out his specific criticisms, which might include dishonesty, disrespect of patients or staff, self-centeredness, arrogance, hostility, indifference to the welfare of patients, lack of interest in medicine, inefficiency, delinquency in completing medical records, or laziness. Often he offered to counsel the young physician or recommended psychiatric evaluation. Occasionally he advised the house officer to switch to a less demanding residency than internal medicine. Rarely he recommended that the individual leave Vanderbilt.

He wrote one such letter to an intern who had exhibited many deficiencies:

> Dear X,
>
> I write you because I think that you (or we) need to make some realistic plans for your future. You have now completed your internship year—your performance as an intern was bad ... You should know that your performance was <u>far</u> worse than that of <u>any</u> other intern in Medicine I have known in ten years at Vanderbilt.
>
> Your histories have often been ... sketchy and superficial. Much of what we learn is from our patients rather than books. You are losing out ... on this source of education, and thus it is not surprising that your knowledge and your clinical judgment remain deficient ... You tend to indulge in ... oversimplification of complex problems, perhaps because you just don't feel like taking the trouble and effort to try to unravel them. Worst of all, you have seemed indifferent to your patients' complaints ... you refused repeated pleas by the nursing staff to come see your patients at night, and another intern would have to be called. You have behaved as if you are ... self-centered—this is a lethal defect in a doctor. The combination of not caring, being ignorant, and being self-centered has made you singularly ineffective.
>
> One simple ... consequence of your performance is that I don't think you will be able to obtain certification from the American Board of Internal Medicine unless you change in a fundamental way ... A more important consequence is that I think you will be unhappy as a doctor, for I believe that happiness comes from doing well whatever it is you choose to do with your life. I suspect that being at Vanderbilt has been an unhappy and

lonely experience ... for your attitudes are so different from those of all of your peers. As you might guess, it has made me unhappy to have you at Vanderbilt, and I hate having to write this letter to you, so the year has been a disaster for us both!

The preceding remarks become of some significance because there are several courses of action open to you. You might ... get out of medicine altogether—I don't think this would diminish your stature one iota, the dignity of any job comes principally from how well it is done, I have never been confident that being a physician is a bit more laudable than writing advertising copy or selling shoes or whatever. You can go to another hospital to do your house staff training—perhaps you are right that the 35-40 hour week is the best way to train doctors, but that does not happen to be the system here (and you were fully aware of that before you came) ... You can enter a part of medicine that will make less personal demands on your physical strength and on your willingness to give—wouldn't dermatology, pathology, radiology be specialties to consider? Or you can try to do a proper job in internal medicine at Vanderbilt—this will require motivation, self-discipline, hard work, honesty (especially with yourself), putting the patient's welfare before your own ... I would prefer that you take this last course of action, but I think that the others are just as reasonable...

Your internship performance ... does not preclude your becoming a great physician ... You have so many important assets—your good intellectual capacity, good education ... and most especially your excellent kind personality. You must decide <u>now</u> (in my opinion) whether you want to use these assets for internal medicine or for some other endeavor. Being an internist brings wonderful privileges with it, but it entails at least an equal amount of responsibility. Do you want this responsibility?

Yours,
Tom Brittingham[183]

T.E.B. AS TEACHER

Brittingham's teaching methods created a special esprit de corps. Medical students felt that they and T.E.B. were all in it together. "We had the same objectives, we were colleagues," said Dr. Liz Kruger. "That's why our group was always driving to Columbia, Fayetteville, and other towns to review hospital records."[184]

The age-old template of the medical history—chief complaint, history of present illness, past history, personal and social history, and review of systems—provided a tried and true system for medical decision-making. T.E.B. highlighted the importance of the personal and social history. These parts of the history were often poorly acquired, even omitted, and were under-appreciated as diagnostic tools by most physicians. To illustrate the importance of the personal and social history, T.E.B. assigned six medical students to take histories of six different hospitalized patients and then present the cases to him. After their presentations he announced that each of the students had missed a critical element of the history: each patient had undergone a traumatic emotional event within the year preceding the physical illness. He sent the students back to their patients to take a more complete history.

T.E.B. maintained that prior emotional stress often preceded a physical illness, and told his students that they wouldn't read this in books. "If you really listen to your patients, they know a lot more about their disease than you do."

Brittingham belabored the importance of directing medical reading to the specific diseases of specific patients. In a conference, however, he rarely quoted the medical literature, preferring to concentrate on the details of the patient. Whenever he did draw attention to a specific journal article, he did it in dramatic fashion. He might stop talking, turn to the bound journals on the shelves in the Julia Weld Conference Room, and pick a volume. He laid the book on the table and opened it—always to the exact page containing the article he planned to cite. Since no bookmark or other device was visible, no one knew how the journal always opened to the appropriate page

While third year students on their Medicine rotation, Robert Faber and David Fleisher were friends and good-natured competitors. One Friday afternoon, after polishing his presentation for T.E.B. the following morning, Faber boasted to his classmate, "Fleisher, I'm going to cream your ass tomorrow." That afternoon Faber drove to Hendersonville, Tennessee, about twenty miles north of Nashville, to review his patient's old hospital records. He assumed this would ice the cake with Brittingham and secure a high grade. At the case presentation Saturday morning, Faber considered his presentation detailed and brilliant. Brittingham seemed impressed.

Then Fleisher presented his case: a homeless man admitted the preceding evening with pulmonary embolism. He had no old records and only a sketchy history. Across the table, Faber began to smile and smirk under his breath regarding Fleisher's substandard presentation. He was sure he had *creamed* Fleisher.

Fleisher had admitted his patient to the obstetrics and gynecology resident's call room (known as the *goat room* for its distinctive odor), because all the regular rooms in the hospital were filled. There were no nurses assigned to the *goat room*. Fleisher had stayed up all night with his patient, whose vital signs were unstable, doing both physician and nursing duties.

When Fleisher finished his presentation, Brittingham paused, then said to the students, "Now let me tell you about a good doctor." Faber leaned back in his chair, put his hands behind his head, smiled, and prepared to receive Brittingham's accolades. He was flabbergasted when Brittingham began to praise Fleisher.

Brittingham pointed out that Fleisher had put his patient's medical care above his own grade. Faber, of course, knew that Brittingham was right—that Fleisher's motive was pure and that his own was self-serving.[185]

T.E.B. did not hesitate to embarrass a student if he felt it was necessary, but he always praised examples of a student's exceptionalism. If a student had researched remote possibilities and shed important light on a case, T.E.B. never forgot it.

Charlie Mayes was working on ward C3100 as a medical student when a pregnant woman with a fever was assigned to him. She was coughing up yellow sputum. Mayes made a gram stain of the sputum to look for bacteria. Seeing none, he then did a special acid-fast stain, looking for tuberculosis. Peering through his microscope, he saw the slide was teeming with the reddish organisms, characteristic of tuberculosis. He had made the diagnosis. When T.E.B. heard of this feat, he made every student come and look at the slide. "I was a hero for a few weeks," Mayes said, fifty years later.

"Dr. Liddle asked me to be chief resident in 1970-1971. I had two kids, a family, I didn't know whether I could afford that. But I decided that to spend time with T.E.B. every day for a year, you couldn't put a price on that," said Mayes. "I loved that year. I got to know him ... he has a guard up ... he didn't want you to know too much about him ... Sometimes he'd invite me to his house ... He would talk about medicine, about life. He loved sports, we'd talk about basketball games."

After completing a cardiology fellowship in Denver, Mayes entered private practice in Charlotte, North Carolina. "I got big time sick in Charlotte, the doctors couldn't agree on what I had. My wife Carol called T.E.B ... He offered to come to Charlotte, he talked to some of my doctors, he regularly talked to me on the phone. I had a rare systemic mycoplasma infection which had infected my muscles, but after three months I got well," he said.[186]

John Schimmel, Ed Anderson, Dwight McKinney and John Dixon comprised another group of medical students who would do anything to curry favor with T.E.B.. Dixon tells the story of the henhouse near Hohenwald, Tennessee. One Saturday morning, Schimmel was to present the case of a farmer with histoplasmosis, presumably acquired in the poultry house. Droppings from chickens, pigeons, starlings, blackbirds, and bats support the growth of histoplasmosis. For that reason, it's particularly common in chicken coops and old barns. Schimmel wanted a polaroid picture of the chicken house in his presentation. "T.E.B. was the happiest man on the planet when he saw that polaroid photo," said Dixon.[187]

T.E.B.'s practice of going to great lengths, such as driving to out-of-town hospitals for medical records, was a metaphor for diligence. His goal was not to one-up or embarrass his students; he was compelled to show what they could learn by thoroughness. "You don't have to be smart to be a doctor, but you do have to be 100% conscientious," he said.

Dr. Gary Hoffman, a Vanderbilt medical school graduate, remembers the "dog story." He said T.E.B. was always telling students and residents that they were "seven times smarter" than he was. One day one of the residents was in T.E.B.'s office and noticed a picture of T.E.B.'s dog. Pointing at the dog, Brittingham said, "That dog is seven times smarter than I am."[188] Nobody knew how to interpret this.

CONFERENCES

T.E.B. always did his homework for a conference. He was going to work harder than anyone in the room. He began each conference by saying, "I'm just here to learn." On occasion he abruptly jumped up from the conference table and ran across the hall to his office, where he kept 3" x 5" index cards of all the patients he had seen. His office also contained multiple tall, disheveled stacks of articles torn from medical journals. There was no obvious filing system, but he knew by instinct where each article lay. He returned triumphantly with an index card or an article relevant to the discussion.

T.E.B'S DEATH CONFERENCES

T.E.B. stressed learning from mistakes. He felt the autopsy was the most important tool to learn what could have been done differently. Mortality conferences at other medical centers had ominous reputations because they were implicit indictments of the medical care that had been provided. After all, the patient had died. The house officers were uneasy because they were held accountable for their actions. No one wanted to be identified as a lesser physician. But T.E.B.'s death conferences were different: The doctors on the case were never named, only identified if they volunteered information. The point was not what one did, but why the decision was made. T.E.B. didn't think he could teach residents how to be better doctors by flogging them.

Despite the so-called anonymous environment, the conference could intimidate the resident involved in the case. Dr. Lawrence Wolfe, the first chief resident to serve under T.E.B., remembers a man with chest pain who was brought by his family to the Vanderbilt Emergency Room. Suspecting a probable heart attack, the resident tried to admit him to the hospital, but all the hospital beds were full. When it was determined that the patient was a veteran, the resident instructed the family to take him across the street to the VA Hospital. Shortly after arrival there, the patient died.

T.E.B. talked to the family later and learned they had no vehicle. They had picked up the patient by the shoulders and dragged him across the street and through the VA parking lot. T.E.B. was furious that no one had arranged appropriate transportation for the patient. During the conference he never identified the resident. But the poor resident sat sweating with embarrassment while T.E.B. admonished him—unidentified.

Dr. Wolfe elaborated, "T.E.B. was a master at high theater. His description of the man being dragged across the street would bring tears to your eyes. That was one of the criticisms of him, that he was so persuasive. He was so good at it that some faculty members objected, because anything that came out of his mouth was accepted as gospel truth by the students and house staff. It was hard to convince them that T.E.B. was ever wrong."[189]

Dr. Robert Dunkerly describes a death conference when he was the brunt of T.E.B.'s attention. As a medical student, Dunkerly had been the Founder's Medalist and had been honored with the Albert Weinstein Prize, the award given to the fourth-year student who had demonstrated high scholastic attainment and qualities which characterize the fine physician. When he was an intern, he saw a 26-year-old African-American woman in the emergency room who complained of chest pain. Examination of her heart and an electrocardiogram were normal. However, she flinched with tenderness when he pressed at the sites where the left second and third ribs joined the breastbone. Dunkerly thought her symptoms represented Tietze's Syndrome, a non-cardiac condition due to inflammation of the joints where the ribs join the sternum. The diagnosis is based upon the ability to reproduce the pain by palpation of the areas of tenderness. (Chest wall tenderness, however, is common and does not entirely exclude the possibility of serious coronary heart disease. Also, electrocardiograms can be normal in patients with coronary heart disease.) Dunkerly prescribed Darvon Compound, a medication then used for low-grade pain.

At a later date the patient came back to the emergency room complaining of similar chest pain. This time she was seen by another intern and thought to have anxiety. She received an intravenous injection of Valium. Her brother, considered disruptive by the emergency room personnel, insisted on further evaluation. This time an electrocardiogram showed an abnormal elevation of the ST segments, the classic sign of a heart attack. As the intensive care unit (ICU) was full, the admitting resident placed her in the ward C-3100 hall, just outside the ICU, with a cardiac monitor. No one was attentive to the monitor. In the middle of the night, she had a cardiac arrest and died.

At the death conference, T.E.B. was simmering: "One of our best residents, he's smarter than I, winner of the Founder's Medal and the Weinstein Prize, thought she had Tietze's Syndrome." T.E.B. told Dunkerly that Tietze's Syndrome certainly wasn't the correct diagnosis, and he'd never seen a patient with Tietze's Syndrome die. He continued, chuckling, "Another of our best residents, certainly smarter than I, gave her intravenous Valium for chest pain."

He saved his final admonishment for the admitting resident. T.E.B. discovered that the average age of the patients in the ICU that night was above eighty. He said the resident should have moved one of the octogenarians out of the ICU to make room for the 26-year-old with a heart attack. "Another of our finest residents put her out in the hall with a monitor that no one watched, and she died unattended. If he couldn't find anyone to watch the monitor, he should have sat there and watched it himself."[190] In both of these death conferences T.E.B. emphasized a doctor's responsibility for the patient, a responsibility that went beyond making a diagnosis and prescribing treatment.

Dr. Sergent told me that the death conferences taught responsibility, but usually someone had made a mistake—and that person knew it. Today's conferences are oriented more toward refining operational systems, organized procedures to use in specified situations—for example, setting the accepted procedure to follow when the ICU is fully occupied. "If all you stressed was someone's mistake, that person would not make that mistake again, but the system wouldn't change," he said. "Now we step back and see how we can create a better system so that this won't happen again—so nobody will make this mistake. I think that's an improvement."[191]

CLINICAL-PATHOLOGICAL CONFERENCES (CPC'S)

In T.E.B.'s day, clinical-pathological conferences were held in an ancient amphitheater with steeply-rising rows of seats. A house officer presented the case of a patient who had died. Next, a faculty member discussed the case and tried to explain why the patient died. Finally, at the moment of truth, a pathologist showed the autopsy findings, establishing the diagnosis and the cause of death. It was designed as a game, but often resulted in an unforgettable learning experience. An example follows:

> The clinical details of the case were obscure. An elderly man was admitted to the VA Hospital. He was alone, no family member accompanied him. None of the doctors knew how he got there. The patient was confused and could relate no symptoms or details of his illness. There were no clues in the laboratory tests, nothing distinctly abnormal on the physical exam, he had no fever, everything was nondescript. After a few days he just dwindled away and died.

T.E.B. was the faculty member scheduled to discuss the case. At first he appeared perplexed. "Here was this old guy, he had chronic brain syndrome (the term for dementia in those days), it was impossible to grab onto anything to construct a diagnosis." He chuckled with his gravelly voice, then said, "Well, I found his brother's phone number. The brother didn't have the greatest memory in the world either, but he remembered that the patient had once been in the old VA Hospital, out on White

Bridge Road. He had stayed there for months, the brother didn't know why. He didn't remember if anyone in the family had visited."

The old VA Hospital on White Bridge Road no longer existed, but T.E.B. knew the old records would be on microfilm somewhere. He found them in a dusty repository and reviewed them on an old microfilm reader. He learned the patient had been treated for tuberculosis about twenty years ago. T.E.B. went back to the recent hospital chart but found only nonspecific abnormalities. When he looked at the chest X-ray, it appeared clear at first glance, but there were scattered little dots in both lungs. T.E.B. decided the chest X-ray finding was compatible with miliary tuberculosis. (Miliary refers to the small dots on the chest X-ray which resemble scattered millet seeds and are a sign of tuberculosis which has spread throughout the body. Untreated miliary tuberculosis is invariably fatal.)

When the pathologist showed the results of the autopsy, he confirmed that T.E.B. had nailed the diagnosis. The old man had died of a curable disease. If his doctors had had the same tenacious determination to track down old information, they probably could have saved his life.[192]

The patient presented at another CPC had died of multiple pulmonary emboli. There was no obvious source of the blood clots which had lodged in her lungs. T.E.B. went to the patient's home town to review her medical records at the local hospital. Afterwards he rested briefly on a park bench at the courthouse square. He asked passersby if they knew the woman who had died at Vanderbilt. One of them replied, "Oh, the lady who swallowed the toothpick?"

At the CPC, T.E.B. discussed the case, postulating that the patient's death was related somehow to the swallowed toothpick. Dr. Robert Collins presented the autopsy findings and demonstrated the patient's heart. The heart contained a toothpick embedded in the posterior wall of the right atrium. The toothpick had migrated anteriorly from the esophagus through the thin wall of this part of the heart. A blood clot attached to the toothpick was the obvious source of the emboli to the lungs. T.E.B. was right again, though this time he had help. After that, the pathologists who ran the CPC's insisted that whenever T.E.B. was the discussant, he had to stay in town. That rule was just for him.[193]

T.E.B.

MEDICAL GRAND ROUNDS

When Dr. David Gregory was a second-year medical resident, he had a middle-aged woman from Hickman County under his care. She had vague symptoms—had lost weight, just didn't feel well. X-rays revealed a narrowed segment in a ureter. Since she had a positive tuberculin skin test, the doctors considered tuberculosis of the urinary tract as a likely diagnosis. Gregory started treatment with INH—isoniazid—a recently available and highly effective drug against tuberculosis. The patient developed spiking fevers and drenching sweats. A liver scan showed a possible abscess in the liver, leading Gregory to ask for a surgical consultation.

Dr. H. William Scott, chief of surgery at Vanderbilt, pronounced that this woman needed the ultimate diagnostic test (at the time), surgical exploration of the abdomen. The evening before the scheduled operation, a light kindled in Gregory's brain: maybe this was a drug-induced fever. He stopped INH, and the next day the patient's temperature was normal. Risking the wrath of Dr. Scott, Gregory canceled the surgery. He waited a few days, then gave the patient one-sixth the usual dose of INH. She had a prompt rigor, and her temperature rocketed to 103. The next day her temperature was normal. Now Gregory was convinced that his patient had a drug fever. He had made her "sick as a dog" with INH.

T.E.B. discussed this case at Medical Grand Rounds. A nurse wheeled the patient into the front of the amphitheater. The packed audience of students, house staff, and faculty peered down as Gregory presented the history. T.E.B.'s premise was that INH, a drug widely used to treat tuberculosis, occasionally caused fever or severe liver damage. He discussed brilliantly how well-meaning doctors sometimes made people sick with drugs. As he concluded, he said, "Well, I've got a positive tuberculin skin test myself," holding up his arm so everyone could see the prominent red spot on his forearm. "You smart doctors all know that when you have a positive tuberculin skin test you're supposed to take INH for a year to prevent active TB. So I went to the pharmacy and got this bottle of INH." He raised the bottle of pills.

He beckoned to a nurse, who brought to the podium a medication tray containing a small paper cup with a single INH tablet. T.E.B. took the tablet in his fingers, held it aloft for the assembled doctors to see, and placed it to his lips. Then he stopped. He stared at the audience for an interminable time, then shook his head and replaced the tablet into the cup. He walked to the corner of the amphitheater, took aim, and threw the cup into the trashcan. With a chuckle, he proclaimed, "I don't think I'll take this pill," and sat down in the front row.[194]

A few months later, T.E.B. spoke at Grand Rounds again. This time his topic was allergy to quinidine, a drug used to treat cardiac arrhythmias. He said that quinidine could cause destruction of the platelets in people allergic to the drug. After he stated his argument, the same nurse brought a similar medication tray, containing a similar small white cup with a single tablet. T.E.B. put the tablet in his mouth, swallowed it, and asked the nurse for a glass of water. "I'm not allergic to quinidine," he joked to the audience, amid raucous laughter and applause.

THE SHOVEL AWARD

The Cadaver Ball is an annual event put on by the graduating class, to celebrate the first-year medical students' completion of gross anatomy. A rowdy affair, it is characterized by heavy drinking, drunken skits, and other displays of unprofessional behavior. In the 1960's the Cadaver Ball was so wild that only one place in Nashville would host it —the old Noel Hotel, which closed long ago.

The Shovel Award was originally awarded at the Cadaver Ball to the faculty member who was felt to shovel bovine feces to medical students the fastest. As time passed, the annual awardee was regarded as the faculty member who was *most influential in the lives and education of the medical students*. Take your choice. The earlier shovels had a shiny chrome flat blade and handle, resembling the type of shovel used by farmers in their barns. The shovels awarded today have a fancy spade-shaped blade. Four gleaming shovels, with engraved names of all past awardees, are displayed in the lobby of Light Hall near the office of the dean of the medical school.

In 1968, the year Neil Armstrong walked on the moon, the students selected Dr. Brittingham as the recipient of the Shovel Award. Brittingham, no seeker of honors or recognition, decided not to appear at the Cadaver Ball to accept the award. He always made it clear that he considered awards, honors, grades, and medical societies just vehicles for the egotistical to pat themselves on their backs. When Dr. John Chapman, serving his first year as dean of the medical school, arrived at the Ball, he learned of Brittingham's decision. The students said, "What are we going to do? We don't want to give the award to him in absentia."

Dean Chapmen called T.E.B. and told him he had an obligation to be at the Noel Hotel in ten minutes. T.E.B. refused.

Chapman pressed on, "Tom, you've got to come, or I'm coming to get you." When T.E.B. refused again, Chapman left the party, got in his car, and drove out to Brittingham's house on Curtiswood Lane.

Once inside, he pressed his argument again, but Brittingham was adamant. Finally, Dean Chapman said, "You can't let those students down."

That statement got to T.E.B. He cocked his head, stood in silence for a moment, then said, "Let's go." He would never disappoint a student in any way. Chapman drove him to the Cadaver Ball, where he received the Shovel Award.

Dean Chapman said, "Brittingham was honored in a way that didn't show."[195]

The Shovel Award was the only award T.E.B. received during his seventeen years at Vanderbilt, or at any other time during his career. The satisfaction of work well done gave him all the gratification he needed.

Photo by Dr. Eric Dyer
T.E.B. teaching at Vanderbilt

12

The Doctoring Professor

> *"Two or three minutes sitting at the bedside is probably worth ten minutes standing at the bedside. Sit down, lean forward, and convey in that window of what Clif Cleaveland calls 'a sacred space between the doctor and the patient' that the patient has your undivided attention."* —Dr. Charles Bryan[196]

T.E.B. never carried a beeper during his his seventeen years at Vanderbilt. However, he believed there was no excuse—plans, fatigue, or inconvenience—that justified not coming to see a patient who needed him. He doctored many faculty members and their families. He was Chancellor Alexander Heard's personal physician, and he treated Vanderbilt's first Nobel laureate, Dr. Earl Sutherland. But every Friday afternoon T.E.B. practiced medicine in a clinic at Nashville General Hospital, the city hospital for poor people. He appeared precisely at 1 p.m., carrying his black bag and vintage Smith Corona portable typewriter.

"His care and compassion for the people he treated at Nashville General Hospital influenced me," said Dr. David Gregory. "Except for his Brooks Brothers clothes, there were no outward signs of affluence. He felt comfortable with poor people and cared for them as skillfully and as thoroughly as if he were taking care of the chancellor."[197]

Brittingham had an unconventional clinic practice for an academic physician. He visited many patients in their homes. He used B-12 injections liberally as harmless placebos. He treated patients who complained of weakness or fatigue with a small dose of thyroid hormone. He called it a therapeutic trial. Many of his patients debilitated by rheumatoid arthritis received only salicylate (aspirin) therapy. On the rare occasions when he left Nashville, he left a telephone number where he could be reached and invited patients to call him collect.

T.E.B.'s handwriting was illegible. Only when he used his typewriter could one decipher his notes. His clinic notes were short. They contained the patient's chief complaint, blood pressure, a brief history, and exam. They closed with a numerical dollar amount—how much money he had given the patient for medicines, groceries, glasses, rent, transportation, and other essentials. Occasionally, a patient would arrive in the emergency room intoxicated a few hours after a clinic visit with the generous professor.

Elizabeth Marlene Hoppe, the unit clerk in the medicine clinic, described Dr. Brittingham as the "sweetest, nicest thing you could ever know." She remembered him always wearing an Oxford cloth shirt, tie, and charcoal gray slacks, never a white coat. His patients loved him because he was so kind to them and got down to their level. He never took a break, never had a Coke or coffee, and seldom left before 5 p.m. She recalled that once he wrote a prescription for Valium and the patient went straight to the parking lot and sold the prescription.[198]

Dr. Mark Averbuch was a medical resident in the clinic. He remembered a patient in whom he had diagnosed a syphilis infection of the aorta. He prescribed a course of penicillin, which cured the problem. On a return visit to the clinic, the patient surprised Dr. Averbuch by requesting a transfer to Dr. Brittingham's care. This request puzzled Averbuch, and he asked the patient why he wanted to switch. The patient responded candidly, "Because Dr. Brittingham gives us money, and you don't."[199] On another occasion, T.E.B. gave a patient $200 after saying, "I know what's wrong with you. You just don't have enough money."[200]

When a medical student or house officer became ill, T.E.B. was usually the doctor. Dr. Dick Dixon developed a viral syndrome when he was an intern. His resident told him to see T.E.B. The professor pricked his finger, did a complete blood count himself, and saw atypical lymphocytes on the blood smear. (Atypical lymphocytes on a blood smear usually indicate mononucleosis. Often a doctor orders a mono spot test to confirm the diagnosis.) T.E.B. said he thought Dixon could work. Dixon asked if he should get a mono spot test. T.E.B. laughed and said, 'Dick, we can tell it's mono without a mono spot test.'"[201]

Medical student Ed Holleran rode his bicycle to Vanderbilt Hospital every day. One morning while bicycling past the VA Hospital, he developed a severe headache and had difficulty pedaling further. Somehow he found his way to Dean Chapman's office. The secretary interrupted Chapman's meeting with a faculty member and brought Holleran, now dizzy and stumbling, into the office. Chapman realized the seriousness of Holleran's condition and immediately took him to T.E.B. After a quick look into Holleran's retinae, T.E.B. diagnosed a ruptured cerebral aneurysm. He called a neurosurgeon, who repaired the aneurysm.[202]

Liz Kruger discovered a breast lump when she was a 23-year-old medical student. She saw a surgeon, who recommended a breast biopsy. He also requested a mammogram before the biopsy. Then, Liz saw T.E.B., who adamantly opposed the mammogram, because he thought the exposure to radiation was unnecessary. He told her to forget the mammogram and just go ahead with the breast biopsy.[203]

On New Year's Eve, 1966, Dr. Gary Duncan, an intern, fell ill with fever and shaking chills. He called T.E.B., who was attending a party. T.E.B. left the party and made a house call in his tuxedo at Duncan's apartment. He sent Duncan to Vanderbilt Hospital, went home to change clothes, and arrived in Duncan's hospital room about 11 p.m. His diagnosis was pneumococcal pneumonia. Duncan felt honored that his health was more important to his mentor than the New Year's Eve Party. He told Brittingham, "I ache all over." Soon a nurse came with an injection that made Duncan "feel wonderful." Duncan discovered later that T.E.B. had prescribed a small dose of morphine.[204]

Dr. Robert Dunkerly described T.E.B.'s care of his mother, who had experienced unsuccessful doctoring from previous physicians. Mrs. Dunkerly was overweight and had severe hypertension. One doctor had instructed her to lose weight and not to return if she didn't comply. Another doctor prescribed two antihypertensive drugs, which caused intolerable side effects. At entry into Vanderbilt Hospital, her blood pressure was 300/160, and she was diagnosed with *malignant hypertension,* a life-threatening condition often accompanied by kidney failure and delirium. Dunkerly called her a "doctor's nightmare" and arranged for T.E.B. to try to help her. She improved a bit in the hospital.

After discharge, T.E.B. came to Mrs. Dunkerly's home every Saturday afternoon to doctor her. He would sit with her patiently and talk her into taking an additional half, or sometimes an additional quarter, of the antihypertensive tablets she was afraid to take. He gradually and cautiously increased the doses. After a few months the blood pressure declined, and the house calls lessened to once monthly. T.E.B. cared for her in this manner for several years.

The foundation of Brittingham's approach was to look for something he could do for the patient. He stressed, "Don't miss a treatable disease." He called hypothyroidism the most important disease in medicine, because specific, inexpensive therapy was available. He emphasized that most people with hypothyroidism look normal. T.E.B. believed that observation of the patient's deep tendon reflexes was a more reliable way to make the diagnosis of hypothyroidism than measurement of thyroid blood tests. He positioned the patient on her knees on the seat of a chair. She faced the back of the chair and held onto the arms of the chair for balance, with her ankles extending behind her. Then, with his reflex hammer, he briskly tapped her achilles tendon, just above the heel. This maneuver demonstrated the prompt contraction but slow, prolonged relaxation of her ankle reflex, the characterlstic sign of hypothyroidism.

Dr. Michael Minch, a surgical resident in the late 1970's, became T.E.B.'s patient when he developed tendonitis. He had taken large doses of over-the-counter anti-inflammatory medications, which led to nausea, vomiting, and headache. Minch described T.E.B. as focused and thorough. He remembered T.E.B. pecking out the history on his typewriter. As he interviewed Minch, T.E.B. asked, "Why would I ask you that question?" Brittingham was always in *teaching mode.*

In a letter to his daughter Susan T.E.B. described his joy in treating patients: "I had a wonderful experience this week. I have a patient named George Riddle, who has <u>terrible</u> crusted skin lesions all over his arms and hands. He has had them for three years, has been to the Mayo Clinic and to Duke as well as to several doctors in Louisville. He then came to Vanderbilt, was seen by all the experts, who said his lesions were self-inflicted. A medical student presented George to me on Saturday morning rounds ... I was positive there was some solution to the problem but thought the odds were about 20:1 against my finding it."

T.E.B. discovered that George's illness began after a cat attacked him savagely when he accidentally killed one of her kittens. The cat lived in an enclosed area surrounded by rose bushes. Soon T.E.B. recognized that George had sporotrichosis. (Sporotrichosis, called "rose gardener's disease" is an infection caused by a fungus. Skin cuts and scrapes that occur when handling sharp-stemmed plants like rose bushes usually introduce the infection. A chronic disease with slow progression, it is difficult to diagnose, as many other diseases cause similar lesions.) George came for weekly appointments and stayed overnight at the Holiday Inn, and T.E.B. got to know him well. George had spent twenty-five years in the Marine Corps and loved barroom fighting. Additionally, he had an excellent ear for classical music. After seven months of treatment, his skin lesions resolved completely. "What a thrill and what an education that was, you just cannot imagine," declared T.E.B.[205]

In the days before the availability of ultrasound and CT scans, the diagnosis of appendicitis was not always straightforward. If undiagnosed, the appendix may rupture, spilling millions of bacteria into the peritoneal cavity and causing peritonitis. One night a resident had abdominal pain and called T.E.B. Brittingham examined the resident, who looked perfectly well and had no fever. He could push his hand all the way to the back of the abdomen without encountering any flinching. The only mark of tenderness was the resident's insistence that it felt tender. With this tenuous evidence, T.E.B. suspected appendicitis and arranged a surgical consultation. The surgeon called ten minutes later and told T.E.B. the patient did not have appendicitis. But at 8 a.m the next morning the surgeon changed his mind and operated. He found an acutely inflamed appendix and removed it successfully.[206]

Dr. Hugh Chaplin and his wife, former next-door neighbors in St. Louis, visited Dotsy and T.E.B. in Nashville one weekend. The phone rang. A patient T.E.B. had inherited from David Rogers was vacationing in North Carolina and was suffering an exacerbation of chronic pain. T.E.B. spoke briefly to her and instead of directing her to the nearest emergency room, he said, "I'll be right there." He excused himself from his houseguests, went immedately to his Thunderbird, and drove all night to Saluda, North Carolina to attend to the patient. He had taken her case from Rogers as a challenge.[207]

Some patients did frustrate T.E.B.. One such patient was a doctor who died from bleeding esophageal varices. (Varices are large, dilated veins in the esophagus caused by cirrhosis of the liver. The fragile veins sometimes rupture, causing massive bleeding and often death.) The doctor's cirrhosis was a result of his overuse of alcohol for many years. T.E.B. had pleaded with him to use marijuana or anything except alcohol. But the patient always replied that only liquor relaxed him. He was aware of his liver problem, but he never stopped drinking. T.E.B. wrote "The doctor had a very active mind and was as smart as anyone could be. The urge to drink must be a powerful one, and I don't know how to get someone to stop. I don't know, I mean I don't know at all."[208]

Occasionally T.E.B. had to hustle. One of his patients was a twenty-seven year old woman, the wife of a medical student and a teacher at Harpeth Hall School. She had acute viral pericarditis, which usually resolves without complications. However, the fluid around her heart was increasing, and T.E.B. hospitalized her. Her husband stayed in the hospital room with her.

The next morning she suddenly felt extremely ill, and her blood pressure fell to 60/50, a value that indicates a very low cardiac output. Just at that time T.E.B. arrived on the floor. With one glance he knew that she was suffering from acute cardiac tamponade and needed immediate pericardiocentesis. (*Pericarditis* is an inflammation of the sac which surrounds the heart. Fluid resulting from the inflammation is called a *pericardial effusion*. If there is rapidly-increasing pericardial effusion, the heart cannot fill and pump blood adequately. This condition is *cardiac tamponade*, and requires emergency treatment. The fluid around the heart can be removed by *pericardiocentesis*, a procedure accomplished by inserting a long needle between the ribs and into the pericardial sac and removing the fluid with a large syringe.)

T.E.B. had never performed a pericardiocentesis. He ran down five flights to the cardiology lab to find a cardiologist to insert a needle into her pericardium. Then he ran back up the stairs. He and a nurse loaded the patient onto a stretcher, and he grabbed the pericardiocentesis tray on standby in the room. In the elevator he thought the patient was going to die. She said she was fading away. She was gray. She could have a cardiac arrest at any moment.

T.E.B. grabbed the syringe and inserted the needle between her ribs into the fluid surrounding her heart. Before he had removed even 50 cc of fluid, she was improved, and by the time 200 cc was in the syringe, she looked better than she had for the past two days. Later that morning she had an operation: The surgeon cut a large opening into her pericardium, so that fluid could not re-accumulate in the pericardial sac. "Pericardial disease is dangerous," T.E.B. said, "but it was fixed by a relatively simple procedure."[209]

T.E.B.'s most reliable partner in doctoring was Dotsy. Dotsy had a knack for counseling women with severe emotional problems. Sometimes these women even lived at the Brittingham home for a while. For example, T.E.B. was treating a doctor's wife from Kentucky. Recently she had a flareup of arthritis and was agitated and depressed. She become wild-eyed and terrified and seemed like a caged animal to her husband. He thought he would have to commit her. Desperate, he called T.E.B. on a Sunday. Without even consulting Dotsy, T.E.B. invited the Kentucky doctor to send his wife to the Brittingham residence, where she would stay in a daughter's bedroom. Dotsy had never met the woman. "The story simply illustrates what faith I have in your mother's ability to understand people, even very disturbed ones, and to make them better," said T.E.B.[210]

On another occasion, T.E.B. received a call at 3:30 a.m. One of the interns had been brought handcuffed to the emergency room and was agitated and combative. The intern was saying "I'm dead." T.E.B. went to the emergency room immediately. The intern said he had had some trouble with his girlfriend. Within an hour he calmed down and was rational. T.E.B. thought the intern needed a protected environment, at least temporarily. So he sent him to the Brittingham home and told him to introduce himself to Dotsy. Dotsy cancelled a scheduled trip to St. Louis in order to watch him. The intern stayed with Dotsy at T.E.B.'s home for two days, then, acting normally, went back to work. T.E.B. wrote, "She did this so well, I knew she would. She is just a fantastic talent."[211]

When T.E.B. left Vanderbilt in 1980, his physician protégés in the Nashville General Clinic assumed the care of his patients. Soon after, the clinic nurse informed one of the doctors that Dr. Brittingham made a monthly visit to a homebound patient and delivered a supply of medications from the closet of drug samples. The doctor drove to the patient's home in a nearby public housing project. A frail elderly woman opened the door and graciously accepted the medicines. Then, with a puzzled look on her face, she peered around the legs of the earnest young physician. "Dr. Brittingham always brought groceries too," she said.

13

Recognition of Functional Disease

"I heard him say, 'Before you call a patient a crock, you have to exclude every possibility, and not many of us are smart enough to do that.'"—Dr. Dan Canale

Crock is a disparaging slang term used by insensitive physicians to characterize patients who have no discernible physical cause to explain their symptoms. Labeling a patient a *crock* implies that the disease process is all in the patient's head. A disease that stems from a patient's emotions is an accepted concept in medicine. Some of the diagnoses which may overlap include functional disease, conversion reaction, psychosomatic disease, psychoneurosis, psychogenic disorder, and hysteria or hysterical reaction.

T.E.B. never used the work *crock* to characterize a patient nor allowed its use by a student or resident. When he was convinced that the patient had a functional disease, he might say something like, "This patient may be difficult to help," but that didn't mean he wouldn't try.

In psychiatry, *conversion* of psychological stress into physical symptoms, a process called *somatization*, accounts for functional disease. Sigmund Freud used psychoanalysis to diagnose and treat many such patients. Freud used the controversial term *hysteria* to label the disorder, because he believed that it was specific to women.

A significant proportion of patients in primary care and many medical specialties have functional disorders. During the period that Dr. Brittingham worked at Vanderbilt, such a patient occasionally entered the hospital as a diagnostic dilemma. Procedures which likely would have helped make a correct diagnosis, such as CT scans, MRI scans, ultrasound, endoscopy, and sophisticated laboratory tests were not yet available. Doctors frequently asked T.E.B. to consult on such difficult patients because he was known to spare no efforts to exclude a diagnosis of functional disease.

One can best understand Brittingham's thought processes about this subject from his own words. He delivered the following lecture to medical students in 1964.[212] I transcribed the lecture from T.E.B.'s faded typewritten notes 52 years later. Parentheses designate my clarifications or comments. "..." indicates sites where I omitted brief sections, for brevity.

> The title of this afternoon's clinic is *The Recognition of Functional Disease*. It is about *crocks*. What I mean by *crock* is a patient whose symptoms ... have resulted entirely from emotional problems and have no basis in disease or dysfunction of any organ.
>
> The reason I am talking about the recognition of functional disease is that it interests and excites me ... The recognition of *crocks* interests me because they are said to be so common. The current edition of Cecil's Textbook of Medicine has a long chapter on the psychoneuroses. It says ... 'an estimated 60% of all patients seen by physicians suffer from some sort of psychoneurotic disorder.'[213]
>
> Making a diagnosis of so-called *crock* also fascinates me because how to do so is so poorly described in all of the (medical) books which I have read. I urge each one of you to read *Cecil* tonight and see how little it says about how to recognize the disease which is going to be many, many times more common than anything else you see. *Cecil* begins with, 'Ideally, the diagnosis of psychoneurosis is based on a comprehensive history-taking and a survey of the patient's patterns of personality development, beginning with the experiences of early childhood.' It really says very little more than this.
>
> Do you know how to make a diagnosis of psychoneurosis now? Stop and think about exactly how

you would do it so that you can explain it to me. Do you think the chances are that the final examination in medicine will contain a single question about how to diagnose a disease which 60% of your patients are said to have? The reason the chances are so poor is that I do not believe there is a single person in the Department of Medicine who could write a sensible answer to that question, an answer which your own answers could be graded against. Are you sure that the commonest of all differential diagnoses you will face, that between structure and function, is easy?

The diagnosis of symptoms caused by psychoneurosis is made frequently. One problem with the diagnosis is that it has certain attractions for the physician which might tempt him to diagnose it when it is not really there. Some can be tabulated.

First, the diagnosis of psychoneurosis can be invoked to account for the symptoms in any patient you see—there is simply no set of symptoms incompatible with it ... A prominent English psychiatrist recently noted that the more slender and insecure is the practitioner's knowledge of disease, the more prone is he to regard strange or puzzling cases as instances of hysteria (psychoneurosis).

A second beauty of the diagnosis of psychoneurosis is that it can never be disproven ... Thus, when the patient I have been following with headaches of psychogenic origin turns out to have a brain tumor, I say that the patient had a brain tumor with severe functional overlay. My diagnosis was right all along, and wasn't she lucky she had psychoneurosis because it probably enabled us to pick up her tumor earlier than we otherwise would have.

The third attraction of the diagnosis of psychoneurosis is that it seems to shift the onus for the correction of illness from you to the patient. I see a 23-year-old girl with recent onset of abdominal pain and vomiting, examine her carefully, and conclude that she has gastroenteritis.

The next day she feels no better, and this makes me angry and frustrated that she is not getting better as I told her she ought to be. Her physical examination is negative; cursory search turns up evidence of a troubled life; I conclude she is neurotic and over-complaining, that's why she is not getting better. It's her fault for being neurotic, not mine for being stupid and careless. A couple of days later she is dead because of an intestinal obstruction which was diagnosed and corrected too late.

We have said then that the diagnosis of *crock* can be made to account for any set of symptoms you are going to see, it can never be disproven, and it seems to take the responsibility for cure of a difficult situation off you and to put it on the patient. Don't you think the diagnosis might be made more often than it ought to be? Perhaps we should be careful about how we make the diagnosis.

What criteria do you use for recognizing symptoms which are purely psychogenic? Physicians almost never state the criteria they have used in making a diagnosis of symptoms occurring on a neurotic basis, but, after looking at hundreds of charts containing a diagnosis of psychoneurosis, one is able to guess at the criteria which have been consciously or unconsciously used ... Let us examine them.

The first (misleading) criterion is the failure to find objective evidence of disease in a patient who is complaining. The difficulty here is that our methods for detecting disease are so insensitive. Most of the time all that we recognize clinically are the end stages of disease or the most extremely severe forms, the top of the iceberg. One might easily lose half of his liver cells or half of his kidney function and not show the slightest objective evidence of what had happened.

Patients with cancer are believed to have had their tumors for years before they become clinically recognizable by anyone. A pulmonary embolus often produces symptoms with no changes in chest X-ray, EKG, physical examination, etc. A patient with carcinoma of the pancreas frequently complains of pain—and is labeled neurotic—for months before any abnormality can be found on the most detailed physical examination or in any laboratory test ...

The failure to find objective evidence of disease is most worthless as evidence for the presence of psychoneurosis.

A second (misleading) criterion used by physicians in making a diagnosis of *crock* is evidence that the patient concerned has a troubled life, life stress. There are several difficulties, however, in using this criterion. One is that you ...have no clear idea of how much life stress is present in a control population of patients with structural disease (physical disease) alone, or with no disease.

One reason you probably do not know this is that ... you do not take good personal histories. One of your classmates—Johnston—does seem to have consistent outstanding skill in this area. This is an excerpt from just one of his histories. The patient was a 72-year-old Negro woman who died with uncontrolled diabetes mellitus and a horrible gas-forming cellulitis (a bacterial infection which produces gas in the surrounding tissues, like gas gangrene). Her story in part is as follows. 'She and her husband have lived in the same house for 43 years. Lately a lot of white folks have moved to their neighborhood ... She and her husband are now the only Negroes in the neighborhood. For this reason she wants to sell the house and move. Her husband won't allow it. She says the neighbors are nice, but she feels scared and out of place. She says this even keeps her awake at night. She is frightened. She says she and her husband owe the Metropolitan Government some taxes. They are having financial problems. He doesn't hold down any job, and she doesn't know where he is most of the time.'

T.E.B.

How do you know that significant life stress is not commonplace among patients who are not neurotic? Perhaps you should try getting to know your patients better. Ask your classmate Johnston how to do it.[214]

A second difficulty in using the presence of life stress to make a diagnosis of psychoneurosis is that life stress frequently exacerbates the symptoms of structural (physical) disease. The following will illustrate.

A 54-year-old woman was carefully followed in the outpatient department of a hospital in another city for two years with a diagnosis of hysteria (psychoneurosis). She had many symptoms of bad brain disease. Now she has a huge meningioma (intracranial tumor, usually benign and curable by neurosurgery). She came from Russia, had always been high-strung and emotional. Four years before this patient's craniotomy (surgical opening into the skull) her only daughter married a no-good refugee with resultant tremendous disturbance in the family. The patient began having headaches at this time; she had never really been the same since. Her husband had been invalided (sic) with heart disease, so she supported the family by working as a machine operator. It was noted that the more aggravated she was and the more pressure she was under, the more frequent were her headaches. Two years before admission she heard suddenly of her sister's death with a brain tumor at age 57, and then, almost immediately had a generalized convulsion ... All of her many symptoms increased when she was tense and anxious. One year ago her daughter remarried, to a boy the family liked very much. For the next two months the patient had no headaches, her thinking improved, she became able to read and understand the newspaper headlines. Physicians considered the patient's insight into her condition was so superficial that no therapy other than reassurance was indicated ... When the patient herself was asked what was the matter with her, she said:

'It's nerves ... I have to do all the work, the cleaning, the doing, I just can't take it.' Her sister attributed the patient's condition to 'the stresses she is under at home.'

Be careful ... you can find life stress in anybody if you look hard for it A third (misleading) criterion is that the patient has an ill-defined story, either because he fails to give a clear concise description of his illness or because of the presence of a multitude of symptoms. An ill-defined story, particularly when it results from the presence of multiple symptoms, evokes frustration and hostility in many of your classmates. The production of hostility in the physician is probably not a sound basis for making a diagnosis of *crock*.

Some patients with disease have lower threshold for symptoms than do others. A great many patients have multiple serious diseases present simultaneously, and the more skilled you become the more you will realize this to be so. Life is tough. Your failure to elicit a clear story may be because ... the patient is not very introspective and is a poor observer of himself, or maybe because you are not a skilled enough history taker, or maybe because the patient just does not have a clearly definable story to tell. Thus the history of a man in the early stages of the bone pain of multiple myeloma, having multiple fleeting pain in many different places, will sound vague. (Multiple myeloma is a type of cancer of the bone marrow, that may involve many sites at the same time.) ... It scares me to use the presence of an ill-defined story in making a diagnosis of *crock*.

A fourth piece of (misleading) evidence is that the patient has an abnormal personality, usually manifested by abnormal behavior. One major problem here is that there are multiple causes for abnormal behavior. These include sleep deprivation, starvation, infections of many kinds, body temperature changes, chronic lung disease, severe hypertension, lead poisoning, hypoglycemia, liver disease, abnormalities of serum calcium or magnesium, drugs of many kinds—for example digitalis or isoniazid—Cushing's syndrome, vitamin B-12 deficiency, hypothyroidism, etc. All of these I can easily document with references. There are probably many other similar treatable abnormalities which could make almost any one of you behave in a repulsively crazy manner, although you had no psychoneurosis to begin with.

Each one of these many causes for abnormal behavior must have been excluded by you before you can reasonably infer that a given patient has a basically screwed-up personality. Furthermore, almost any disease of the central nervous system can produce abnormal behavior, and it can occur in any one of you. Therefore, you must exclude all central nervous system disease in the patient whom you want to label as having an abnormal personality. Is this going to be easy for you to do?

Slaton, a British psychiatrist, says that the presence of abnormal personality traits such as attention-seeking, exaggeration of symptoms, dependence, immaturity, indifference ... is irrelevant to making a diagnosis of hysteria (psychoneurosis), that patients lend the stamp of their personality to their symptoms whether they have an organic or neurotic disorder. He might be right ...

Finally, most thoughtful physicians believe that the incidence of serious structural (physical) disease is at least as high in patients with abnormal personalities as it is in persons with healthy ones.

A fifth piece of (misleading) evidence used commonly by physicians in making a diagnosis of psychoneurosis is the presence of anxiety in the patient. But if one is seriously ill, anxiety is a relatively normal response. Thus if one of you—psychologically healthy students—were to have a large pulmonary embolus (a pulmonary embolus is a blood clot which developed elsewhere in the body and traveled to the lung), or to develop pulmonary edema (fluid in the lungs causing serious difficulty in breathing), or were developing tetanus or rabies, or if an embolus suddenly lodges in to your superior mesenteric artery (an artery in the abdomen which supplies blood to the intestines), etc., etc., etc., you will feel and show extreme anxiety even though there may be nothing for your doctor to see, hear, feel, or measure except for your terrible anxiety. If an embolus lodges in your gut, you will have abdominal pain and will be upset and agitated about it, but for many, many hours you will show no abnormality on physical examination or laboratory examinations or all kinds of X-ray exams, yet it is only in these hours in which your disease can still be effectively treated.

A few of you may be devotees of Marcus Aurelius and believe that it is improper to accept death with anything but a blind shrug, whether the death be by being thrown to the lions or by pulmonary edema, and therefore you few may consider it ridiculously neurotic and egocentric to become anxious because one goes into pulmonary edema. The majority of people in the United States today, however, are not students of Marcus Aurelius.

A sixth (misleading) criterion ... is that the patient's symptoms are made better by psychotherapy, for example, by reassurance or by placebo therapy. If the patient responded to psychotherapy, then he must have functional disease. But many physical diseases are self-limited. Failure to die or deteriorate does not prove psychogenesis. More important, the well-educated doctor knows that the placebo is a therapeutic tool of enormous power. Thus the pain of battle casualties or of post-operative patients can be effectively relieved by injections of sterile water ... You all realize, I hope, that it is possible to do almost any kind of surgical procedure ... using hypnosis. This does not prove that the operation performed was imaginary, all in the patient's mind. You might suppose it is *crocks* who respond best to hypnosis, but a recent book on the subject describes the person most likely to respond to a placebo as extroverted, sociable, self-confident, and dominant. This is a *crock*?

How well do you think that patients with conversion reactions (psychoneurosis) respond to simple psychotherapy, to placebos? ... You'll have no courses on this in Vanderbilt Medical School.

So, to use placebo response in making a diagnosis of psychoneurosis is probably idiotic.

A seventh (misleading) criterion for spotting a crock is the presence of symptoms which conform to no recognizable pattern of disease. This is probably an excellent criterion, but sometimes ... an unphysiologic symptom is just a manifestation of technical incompetence in the examining physician. And sometimes unphysiologic symptoms or signs are turned up because one is looking specifically for them. We find what we seek. Thus ... if you will do pinprick examinations after telling the patient to say "yes" when he feels the pin and "no" when he doesn't feel it, you will get some unphysiologic responses in non-neurotic subjects, especially if you encourage these responses a little and the subject is not smart...

An eighth (misleading) criterion used by physicians in making a diagnosis of psychoneurosis is that a good psychiatrist, one whom the physician respects, finds the patient to be hysterical (neurotic). What evidence can the psychiatrist be using that is not perfectly accessible to you? ... This is probably because psychoneuroses are poorly defined, there are no sharp and testable criteria for diagnosing them, the evidence necessary to make such diagnoses is not clear. It is not clear in the psychiatry books any more than it is in *Cecil* ...

So the diagnosis of *crock* is frequently made, but it is often on terribly uncertain grounds, it is a terribly fallible diagnosis. This is fine if the diagnosis leads to wonderful things happening to your patient, but does it? What happens to the doctor's attitude toward the patient after the diagnosis of hysteria or severe psychoneurosis has been made? Is this good for the patient? If the patient really is a *crock*, really has had an unhappy deprived life and has adjusted to it badly, does he probably need more or less tender loving care than the average patient? Does it make any difference in how carefully and alert the doctor follows the sick patient if the doctor's recorded diagnosis is abdominal pain of undetermined origin as opposed to hysteria? Which patient probably gets the more skilled followup?

But you may say that you are going to take wonderful care of people with severe emotional problems by simply recognizing them and then sending them to a psychiatrist. Well, how useful is psychotherapy in conversion reactions?

There is some specific evidence about this. A recent excellent study was done by three psychiatrists from Johns Hopkins. The study was entitled *Contemporary Conversion Reactions*. It was the authors' opinion that they had satisfactorily identified 134 patients with conversion reactions. Their observations on psychotherapy are of interest.

The patients seemed completely refractory to psychotherapy, insisted that their problems were physical and not emotional, attributed anxiety and depression to physical disease, and refused psychotherapy even on a trial basis.

We have used this afternoon's clinic to talk about the (misleading) criteria used in making a diagnosis of *crock*. These criteria have not been easy for me to find; they can certainly be improved upon by you. You should constantly be seeking to develop your own more adequate criteria. I do not think you are going to find them in a book or an article or by talking to your assistant resident.

We have paid our respects to some of the difficulties present in using the available criteria. It is possible that you have come to feel that the diagnosis of conversion reaction or related entities is a diagnosis made only with great difficulty, and that such patients, once diagnosed, are difficult to treat ... You may now be silently skeptical of the diagnosis, knowing what a formidable challenge it is to make such a diagnosis stick over the course of a long-term followup, knowing that you are faced with an even greater therapeutic challenge.

Some of you may come to share my feeling that pure functional disease is very uncommon on a medical ward. In the out-patient department or in practice it may be different, I just don't know. I have one more patient to tell you about. He seems to me pertinent to today's general subject.

It concerns Mr. B., a forty-year-old white male who had become completely disabled by transient severe pains in his legs which occurred hundreds of times daily for several years. He was not working, was not supporting his family, was miserable all of the time. Several faculty members of the Department of Medicine tried to help him in our medical clinic and failed completely.

Because he had a past history of syphilis ... I thought he might have the lightning pains of tabes dorsalis (spinal cord disease occurring in advanced syphilis) and gave him dilantin, with much positive suggestion. (Dilantin and similar anticonvulsant medication are sometimes used to suppress pain from neural sources). This failed completely. He was ultimately referred to the neurology clinic, where a leading neurologist said that his problem was psychoneurotic, and that he of course did not treat that type of thing, but suggested that the patient be sent to psychiatry. To psychiatry the patient went, was diagnosed with hysteria, naturally. Now this was really not a crazy diagnosis, for he had been an alcoholic in the past, and at age 22 he had married a 38-year-old woman, really a bit unusual. So they locked him up on the psychiatry ward at the VA Hospital for more than two months. He received multiple amytal interviews (psychiatric interview after intravenous barbiturate injection) and other forms of psychotherapy, but the last note said that he was never able to accept his leg pains as psychiatric, and he refused to come back to the hospital or to psychiatry because he felt that nothing would be gained by it.

He remained completely disabled for another couple of months, then happened to go to an osteopath in Mt. Juliet (a small nearby town), the only doctor of any kind in that community. The osteopath gave him *furadantin* (an antibiotic) for a urinary tract infection ... Mr. Bennett was astonished to find his leg pains clearing completely on the *furadantin*. He has had trivial recurrences since then, each time cleared by *furadantin*. He does not come back to psychiatry. He does not come back to neurology. He does not come back to the medical clinic, he has no need or desire to do so. His wife says he is doing beautifully. For the past 1 & 1/2 years he has been working regularly at his old and gainful occupation, not bothering any doctors. Now, in my opinion the cause of his leg pains is still indeterminate, whether they were psychogenic or not I do not ever expect to know. This indeterminate state has been the end of many of the so-called functional problems I have seen. It is the stuff of what makes this area of medicine such a tough one to master. Many people think that the best test of a physician is not what he knows to do with a written examination, not how many honors or Boards he has ... but whether he can make sick people get better. Who then was the best physician for Mr. Bennett? Was it me or one of the other members of the Department of Medicine, was it the leading neurologist, was it one of the several psychiatrists who dealt with him, or was it the osteopath in Mt. Juliet?

Thank you.

Brittingham began this lecture by discussing a generally-accepted diagnosis, functional disease, then thoroughly dissected it, persuasively leading his students, step by step, to question its validity—or at least its frequency. He recognized the uncertainty of diagnosis in many patients, and the even greater uncertainty in treating them. Inherently skeptical of authority, he taught his students and house staff to be skeptical and to rely on their own observations. He had doubts about several medical doctrines and did not cower in awe of so-called experts or of publications from prestigious institutions.

In this lecture Brittingham urged his students not to abandon their patients because of lack of a precise diagnosis, but to try to help them, regardless of the challenge. He was more interested in a patient getting well than in making a diagnosis. "Look for something you can treat," was his message. This standpoint is embodied in a quote attributed to Ambroise Paré, a medieval French surgeon: "To cure sometimes, relieve often, and care always."

At the time of this lecture, T.E.B.'s best friend, Dr. David Rogers, was Chief of the Vanderbilt Department of Medicine. Rogers was occasionally asked to consult on a patient suspected to have functional disease. A consultation by the Chief of Medicine may have been an effort by the patient's personal physician to demonstrate to the patient, often an individual prominent in the social or business community, how difficult it was to determine the proper diagnosis and treatment.

Dr. Thomas Frist, a highly-regarded Nashville internist and subsequently the founder of Hospital Corporation of America, hospitalized such a patient and asked Dr. Rogers to consult. This previously healthy woman was now incapacitated with poorly-defined symptoms which conformed to no known disease. Dr. Rogers visited the patient in her hospital room, had a long conversation, and examined her. Subsequently he visited her daily for brief chats. On Saturday morning, while the patient was still hospitalized, Dr. Rogers invited her to accompany him to the Vanderbilt football game. That afternoon the two of them attended the game. A few days later, she felt better than she had in years and was discharged from the hospital. No one ever made a diagnosis.

T.E.B. certainly believed that emotional disturbances could cause physical symptoms. Once a medical resident came to see him because of abdominal pain. Dr. Brittingham obtained an upper GI x-ray exam, which was normal. Then he told the resident that his pain was due to stress caused by his fear of going to Viet Nam. Once the resident heard that, his pain went away.

Dr. Clif Cleaveland reported in his book *Sacred Space—Stories from a Life in Medicine*, that one day during his year as chief medical resident he was unable to work due to a particularly severe migraine. "I thought my head would explode," he wrote. His wife consulted Dr. Brittingham, who asked her to bring Dr. Cleaveland to his office. Following a thorough examination, T.E.B. sought the opinion of Dr. Cleaveland's wife.

"I think he's working too hard," was her reply. T.E.B. agreed with her diagnosis and advised his chief resident to take a couple of days off.

14

Myths, Habits, Beliefs, and the Quirky Thunderbird

"This past week was quite a hard one of work at the hospital, this always cheers me up."—T.E.B.

MYTHS

After T.E.B.'s influence had been felt at Vanderbilt for a few years, exaggerated stories and legends about him circulated, like the one that he worked for a token $1 a year. The truth was that he received both a salary and health insurance from Vanderbilt. The salary was nowhere near what he was worth, but he had little interest in trivialities.

Another myth surrounds the purchase of his and Dotsy's home on Curtiswood Lane in Nashville. According to the story, Dotsy rented a helicopter and hired a pilot to fly her over Nashville so that she could survey the real estate scene. After inspecting the city from the air, she peered down from the helicopter and said she wanted a particular house. When she was informed that she had chosen the Tennessee Governor's Mansion, which was not for sale, she pointed to a house across the street from the governor's mansion and said that one would do. The pilot landed the helicopter nearby, and she notified the occupants of the house that she would like to buy their property. After some discussion, they settled on a price, which Dotsy, according to the story, paid promptly with a check. The homeowners, surprised that anyone could write a check for the entire amount for their large house, called Dr. Brittingham at the hospital in St. Louis to see if the check would clear the bank. He told them the check would clear, but, if they preferred, he would pay with cash.[215]

This makes for a colorful tale, but the only truth in it is that the Brittinghams indeed bought a house at 867 Curtiswood Lane, near but across the street from the governor's mansion. No helicopter was involved. Dotsy did not write a check on the spot. T.E.B. did not offer to pay cash. Actually, Dotsy and her eleven-year old daughter, Susie, traveled from St. Louis to Nashville to look for a new house. T.E.B. stayed at work at St. Louis City Hospital. He and Dotsy discussed the property by phone. Dr. Brittingham didn't see the house until the closing. George Pickens, a Nashville realtor, had shown Dotsy several properties. One of these was the stately home on twenty-five acres along White Bridge Road, previously owned by Dr. Hugh Morgan, the former Chairman of the Department of Medicine at Vanderbilt. Dotsy liked the Morgan house and property, but T.E.B. scotched the deal over the telephone. "The new boy should not live in the professor's house," he said.[216]

Their new house came with woods and meadows, enough space to keep the active Brittingham children occupied. In the seventeen years they lived there, the children spent hours riding horseback over the ten acres and wading up and down the creek which traversed the property. In addition to keeping horses for the children, the family always had a dog. Their youngest daughter, Sally, became a friend of Julie Dunn, Governor Winfield Dunn's daughter. She visited at the Governor's Mansion across the street several times and referred to the governor as "Winfield," but this was at the Brittingham's dinner table, not at the Mansion.[217]

T.E.B. never encouraged the myths about him, but he took no action to dispel them.

HABITS

T.E.B.'s only hobbies were music and books. He owned an extensive collection of records. He favored classical music, especially Bach. He didn't have a second favorite because he considered all others far inferior to Bach. He also liked jazz and Brahms waltzes. His favorite pianist was Dinu Lapatti, a Romanian classical pianist and composer, particularly noted for his interpretations of Chopin, Mozart, and Bach. T.E.B. put on records as soon as he came home, then listened while he worked at his desk in the evening. During one summer weekend while Dotsy and the children were visiting her parents in Long Island, he listened to the blind German organist Helmut Wolcha play Bach, and read a paperback entitled *Four Great Russian Short Novels,* by Turgenev, Dostoyevsky, Tolstoy, and Chekhov.[218] In December,1963, he used the cash Christmas present from his mother to buy several records, including two Haydn quartets, two Mozart quartets, two Beethoven quartets, Beethoven's Archduke trio, Mendelssohn's Scotch symphony, and Vivaldi's Gloria.

When a relative visited Nashville in 1964, T.E.B. and Dotsy took their guest to the Ryman Auditorium for a Saturday night performance of the Grand Ole Opry. T.E.B. enjoyed the performance but never repeated it—despite the fact that Sarah Cannon (Minnie Pearl) was a neighbor on Curtiswood Lane and a close friend of Dotsy's. T.E.B. appreciated the sentiments in the sad country music songs, but he owned no country music records.

He ate dinner with his family every evening about 7 p.m., and when guests were present, even the children's friends, they sat next to T.E.B., and he talked genially with them. He occasionally returned to the hospital afterwards, but usually spent the evening listening to music and reading. First he answered all his correspondence and cleared his desk. If he had received a letter, he replied to it that evening. He had interest in a wide range of books, and Dotsy bought him novels. In 1961 he particularly enjoyed reading *To Kill a Mockingbird.* In later years he admired the author John McPhee. He read medical journals almost every evening. He subscribed to *the New England Journal of Medicine, Blood, Journal of the American Medical Association (JAMA), Annals of Internal Medicine, Archives of Internal Medicine,* and possibly others. Once, when he felt he wasn't reading fast enough, he took a mail order speed-reading course, *Evelyn Woods Speed Reading,* which President John F. Kennedy had used.[219]

Numerous letters to his mother and children disclosed his enjoyment of parties and social dinners, but he also wrote that he usually preferred quiet evenings at home, reading or listening to music. To get his thoughts off medicine, Dotsy dragged him to numerous movies, which he almost always enjoyed, but she said his heart and soul were only with patients and medicine.[220] in 1972, he particularly relished *The Last Picture Show,* a movie depicting a group of high schoolers who came of age in an isolated, West Texas town that was slowly dying, both culturally and economically.[221] Among the hundreds of movies he saw, he commented favorably about *Das Boot*, which portrayed the German submarine service in WWII, *Coal Miner's Daughter,* and *Ragtime.* He did not like *E. T.*

Brittingham gave up tennis when he married. He did calisthenics before bedtime, and the chinning bar was his only other exercise. But he never took an elevator; he bounded up the stairs in the hospital, with students panting to keep up with him. When his faculty friends at Vanderbilt, Drs. David Rogers, Robert Collins, John Flexner, David Law, and Robert Heyssel, invited him to accompany them on camping or hunting trips, T.E.B. replied, "No, I'll leave that to you tough guys." He was not an outdoorsman and had detested camping since his Army days in the Philippines, where he had to sleep on the ground for days. He said he went a year there without a shower.

David Rogers tried repeatedly, usually unsuccessfully, to get T.E.B. to loosen up and enjoy life outside the hospital. Soon after the Brittingham's arrival in Nashville, Rogers and his wife Corky invited them to take a float trip in a canoe down the nearby Narrows of the Harpeth River. Brittingham did not particularly enjoy gazing at the cottonwood trees or the turtles sunning themselves on logs. He did the paddling, while Dotsy read aloud from Erich Fromm's book, *Mans's Search for Meaning.*

The second and last trip down the Narrows took place on a chilly day in 1968. During that trip the Brittingham's canoe capsized, and they began to chill in the wind. T.E.B. decided to build a fire to help them dry out. Unfortunately, he collected a large pile of poison ivy vines for fuel, and stood in the fumes. That night he developed a severe case of poison ivy. This contact dermatitis became complicated by bacterial infection and fever, and T.E.B. had to stay in bed for one of the few times in his life.[222]

He did take his children fishing at a trout farm in Bucksnort, Tennessee, about an hour's drive west on Interstate 40. The trout farm's owner, the son of a radiologist at Northwestern University, had grown up in Chicago but craved the rural life. One Sunday morning the Brittingham family all went to church together, then out to breakfast, then to the trout farm. T.E.B. described the fishing as "excellent." The children had caught 14 fish in less than an hour.[223]

Brittingham took pride in the family aquarium at his home. He described it as murky with algae, so that the baby fish could hide from the monsters. When he added a fiddler crab, he described the crab as "Quick as a cat, one big claw for fighting, one little claw for reaching."[224]

T.E.B. had no desire to travel abroad. He took vacations on the Lambshead Ranch. In 1971 Dotsy and the children vacationed with Vanderbilt Chancellor Alexander Heard and his wife Jean and their children for six weeks on Gooseberry Island in Nova Scotia. T.E.B. visited for two weeks, all he could spare from his duties at Vanderbilt. He also went to annual stockholder meetings of Lumber Industries, the family business started by his grandfather. He expressed his occasional discomfort with the company's operation in a letter to his daughter Susan, "Some day we may be devoid of all our money ... but money should not be what really counts in this world."[225] He characterized scientific meetings as "straight baloney, a scheme to make people feel important." He did not travel to other cities for medical meetings.

T.E.B. never shared how much cash he carried in his wallet. Once he lost his wallet at the Dallas-Fort Worth airport. It contained $1,500, and Brittingham gave its finder a $500 reward.[226] Dotsy told Dr. Eric Dyer that he always paid in cash when they ate dinner at a restaurant. His favorite Nashville restaurant was The Shack, a fish place north of the city. A basket of peanuts on every table there resulted in a floor covered with peanut shells. In Fort Worth, his favorite restaurant was Joe T. Garcia's, where he loved the guacamole, enchiladas, rice, and beans. He was enthusiastic about the Friday Night Fish Buffet at the Rivercrest Country Club.

Though T.E.B. always greeted people enthusiastically, chatted, and chuckled with them, he was a private person. Few medical residents spent any time with him outside the hospital. He and Dotsy did invite groups of residents and their wives to dinner at their home. Once they had a a sit-down dinner for 70, all the residents plus a few of the faculty. They knew everyone's names and greeted each guest graciously. At some point during the evening, T.E.B. disappeared, possibly to the hospital, one never knew.

BELIEFS

When Congress established Medicare and Medicaid in 1965, T.E.B. was enraged that so many doctors and the American Medical Association had fought against the legislation. It was obvious to him that people over 65 needed more medical care and could afford it less than younger people. Brittingham's usual distrust of entrenched physician attitudes led to his feeling that self-righteousness among physicians made them insensitive to the real needs of patients."[227]

In 1967 T.E.B. stated his views about the nation's foreign policy, "I have finally decided that the war in Vietnam is absolutely wrong, represents poor practical judgment but something much worse morally. What a contrast from World War II, where we really seemed to have to fight ... I would refuse to go to Vietnam as a soldier even if ordered to do so."[228] Later he wrote, "The South Vietnamese do not want us there, all we do is destroy them and their country."[229] And in 1973 he wrote that Mr. Nixon seems to be an extremely evil and dangerous man and should be impeached.[230]

He voiced awareness about social inequities and his distressing toleration of them:

> More and more it seems clear that the inequities in wealth in the world cannot be tolerated. Americans use so much energy, live so lavishly, while others have so little. Worse of all is middle Africa, where drouth (sic) and starvation have killed all the animals and are now killing many of the humans. I have a hard time coming to grips with this ... We live in such affluence, even by the standards of wealthy Americans, and gradually I am becoming perfectly accustomed to it and accepting of it. I do need help.[231]

T.E.B.'s religious beliefs seemed to grow over the years. He wrote to his mother in 1961, "I often find it hard to know the will of God, perhaps because I have not yet become a Christian, am still a sideline spectator who manages to attend church once weekly but do little else."[232]

Twelve years later he wrote he believed that all people were equal, they all represented Christ abroad on this earth.[233] After eight more years, when he learned that a resident at Vanderbilt had been verbally abusive to a nurse, he wrote to the resident, "You behaved badly and in a manner not well suited to promote the strengthening of Nursing at Vanderbilt ... and not ideal for advertising your commitment to Christianity."[234]

Dr. Brittingham' kept in his office a framed quotation by Dr. Robert Hutcheson, which served as his guide for the practice of medicine:

> From inability to let well alone
> From too much zeal for the new and contempt for what is old
> From putting knowledge before wisdom, science before art
> and cleverness before common sense,
> From treating patients as cases,
> From making the cure of the disease more grievous than
> the endurance of the same,
> Good Lord, deliver us.[235]

THE QUIRKY THUNDERBIRD

When Dotsy's Aunt Poosie died in 1960, T.E.B. was the recipient of her two-year-old '58 Thunderbird. "The great advantage of old cars is that they teach you how engines work—by their failures," T.E.B. wrote.[236] He drove this car when he lived in St. Louis, the entire seventeen years he lived in Nashville, and until the end of his life in Fort Worth. Before he owned the Thunderbird, he often rode his bicycle from home through Forest Park to work at Barnes Hospital. Dotsy said he didn't like to be seen in the Thunderbird then, because it appeared opulent. Therefore he drove to work early and to home late.

By 1971 the 13-year-old Thunderbird appeared rusted out and had 90,000 miles on its odometer, but T.E.B. didn't like new stuff. The car attracted no special attention when he made house calls in the projects. A boom box filled the back seat, its purpose to replace the non-functioning car radio. Dr. Jean Ballinger was a first-year surgical resident when she developed Hepatitis B. T.E.B. was her doctor and prescribed three weeks of bed rest at home. Convalescing, she walked the few blocks from her home to T.E.B.'s office at Vanderbilt for a checkup. After the clinic visit, T.E.B. insisted on driving her home. The rusty passenger door was stuck, and Dr. Ballinger, a strong woman, could not open it. T.E.B. had to wrestle the door open for her.[237]

The deterioration of the body of the car was progressing so rapidly that he feared the body would drop dead before the engine. He scheduled a paint job and told the workers at Walker Brothers Auto Repair Shop that he didn't care how the car looked, he just wanted the body to last longer.[238] The off-white paint applied didn't appear much different from the old faded silver finish.

One day he noticed a smell of gas, and the needle on the gas gauge dropped rapidly. The gas pump was leaking. Three weeks later the windshield wipers quit. Driving home, he encountered a torrential downpour. He drove home in the dark, without wipers, with his head out the window most of the time. He said he felt limp and old when he got home.

After 15 years and 114,000 miles, the ignition shorted out and caused a fire under the dashboard.[239] T.E.B. feared that the Thunderbird was failing, took it to mechanics at the gas station, then to the Ford dealer, but nobody could find anything wrong. He said he felt sympathetic toward patients who felt ill, went to a doctor, and were told nothing was wrong.

In its 24th year, after 180,000 miles, the Thunderbird showed signs of a failing transmission. When T.E.B. shifted to GO, nothing happened; after a few minutes he heard a loud grinding noise and the car went forward slowly. He could barely get the car out of the driveway. With some doctoring from his mechanic, the transmission problem improved for a while. Then the car had a sudden braking failure in the Harris Hospital parking garage but somehow he controlled the car enough to avoid a crash. T.E.B. arranged for the engine to be rebuilt, confessing, "This may be a monumentally foolish decision, but it has served me so well for so long that I just couldn't face putting it to sleep."[240]

After rebuilding the Thunderbird's engine, T.E.B. learned that he could buy an adapter for the cigarette lighter to power a tape player. He would be able to listen to music in the car again. "That sounded just great," he said, "except that my cigarette lighter has not worked for 15 years."[241]

T.E.B.

T.E.B. in quirky Thunderbird

15

Controversies at Vanderbilt

"It seems cowardly to say nothing when one believes that someone is doing wrong"—T.E.B.

Dr. Brittingham held controversial views regarding some medical procedures and treatment and doubted the existence of several diseases. To name a few examples, he argued against kidney biopsy, chemotherapy for lymphomas, pulmonary artery catheterization, and coronary arteriography, based on his estimate of the risk-benefit ratios. He cautioned against steroid therapy for autoimmune diseases and against insulin treatment of type 2 diabetes. He suspected that Hodgkin's disease might be a manifestation of an undiagnosed infection, such as tuberculosis or histoplasmosis. And he thought polarteritis nodosa, lupus, and sarcoidosis were questionable entities.

T.E.B. liked the word "pithy." In a lecture at the VA Hospital on polyarteritis nodosa, he began, "You expect me to say something pithy about polyarteritis nodosa." (Arteritis is an inflammation of arteries. Polyarteritis is a term which means the inflammation is generalized. The disease is *autoimmune*, meaning that antibodies produced in the body are the cause.) Then he presented a series of cases diagnosed as polyarteritis. One actually resulted from an atrial myxoma (a tumor of the inside of the atrium of the heart), another from a ruptured appendix with an abscess around it, and others from similar unrelated causes. He concluded, "I guess what I'm saying is the diagnosis of polyarteritis, using the argot of today's youth, is a bummer."[242]

T.E.B. also was skeptical that lupus was a discrete disease, because it had so many different manifestations. He irritated his colleagues in the field of autoimmune disorders, who insisted that lupus was a distinct disease. But T.E.B. stuck to his views. He took care of these patients as well as he could, even if he believed that their diagnoses were questionable.

In 1971 T.E.B. gave an informal talk on diabetes to local diabetes specialists. He thought that insulin treatment was not a smart choice for patients with type 2 diabetes, based on articles he had read suggesting that insulin could cause kidney damage from an immune reaction related to insulin antibodies. "The experts sure did not like what I had to say. They seemed rather closed minded to me," he said.[243]

T.E.B. dismissed most of the criteria used by other doctors to diagnose pulmonary emboli. He felt that lung scans were nonspecific, that pulmonary arteriograms were susceptible to uncertain interpretation, and that a laboratory test called LDH was worthless. In fact, he called a journal article from Harvard and the Peter Bent Brigham Hospital supporting the use of the LDH test in the diagnosis of pulmonary emboli as "garbage."[244]

Dr. Roger Des Prez commented that T.E.B. did not fit the usual mold of the upwardly mobile academician. Des Prez said that T.E.B.'s attitude, which was taken by some as heterodox for the sake of heterodoxy and resented as such, actually was part of a religious reverence he felt for real facts. He emphasized that T.E.B. was an absolute master at destroying assumptions masquerading as facts and added that T.E.B.'s unconventional attitudes toward both diagnosis and treatment were more popular with students and house staff than with faculty.[245]

When Agnes Fogo was a medical student, she presented a patient with kidney disease at Professor's Rounds. Dr. Earl Ginn, the chief of nephrology, recommended a kidney biopsy. At that time, performance of a kidney biopsy involved blindly passing a needle through the skin into the expected location of the kidney. Sometimes accidents occurred, primarily bleeding. Kidney biopsies now are guided by ultrasound or CT scan, to assure the needle is in the correct position. T.E.B. adamantly opposed Ginn's recommendation, arguing that kidney biopsy could have complications, and that the diagnosis gained by the biopsy was usually not helpful in arriving at a successful treatment. A vigorous argument occurred, one which Fogo has remembered for years. "I personally am not sure I have ever seen a renal biopsy which had clinical relevance, but that is just a quirk of mine," T.E.B. wrote to a resident.[246]

T.E.B. had heated discussions with his friend, Dr. Robert Collins in

Pathology, about the nature of Hodgkin's disease. Collins was a recognized expert on the pathology of lymphomas and had authored a system of classifying them. Brittingham believed that Hodgkin's disease was an inflammatory process and probably an infection. When he looked at microscopic slides of Hodgkin's Disease, he saw numerous inflammatory cells. He was adamant that none of his patients should be poisoned with chemotherapy.

At a Grand Rounds lecture Brittingham presented a series of patients with Hodgkin's disease. He had treated none of them with chemotherapy, and all were doing well. T.E.B. stated that once when driving through Franklin, Tennessee, he had stopped for gas. The service station attendant told him that he had Hodgkin's Disease in his twenties, and the doctors had prescribed chemotherapy. The family refused this treatment and took the boy home. That was eighteen years previously.[247]

Collins believed that the characteristic cells of Hodgkins's disease, the elusive multi-nucleated Reed-Sternberg cells which looked like an owl's eye through the microscope, were definitely diagnostic of a neoplasm. T.E.B. would have nothing of it. When a young man with Hodgkin's disease died, Collins told his colleague, Dr. Agnes Fogo, that he never saw a person look so heartbroken as Brittingham. T.E.B. had treated him for infection. He finally realized that his treatment hadn't worked, and that maybe he was wrong.[248]

Photo by Dr. Eric Dyer
Dr. Robert Collins

T.E.B.

T.E.B. wrote, "I gave a talk on lymphosarcoma (a malignant tumor of lymphoid tissue) yesterday. I thought it was a great talk, but it was not very well received—people are not crazy about having you tell them that things are different from what they believe, or that you disagree with them. But I loved giving the talk and organizing my thoughts."[249]

What was the basis of T.E.B.'s opinions about hematologic malignancies? He knew that other members of the hematology department at the time were advocating treatment with chemotherapy and radiotherapy, despite the absence of what he considered rigorous studies. He was skeptical that these treatments did any good and thought they made people feel worse. For T.E.B., if chemotherapy didn't work, its side effects were unacceptable. Later research has clarified that chemotherapy is effective in many lymphomas and curative in some.[250] For instance, the groundbreaking MOPP studies at the NIH showed that chemotherapy was often dramatically effective in Hodgkin's disease.[251]

T.E.B.'s greatest challenge at Vanderbilt was the illness of his close friend and colleague, Dr. Glen Koenig. Koenig was a Professor of Medicine and Chief of the Division of Infectious Diseases at Vanderbilt when he became ill in 1969. T.E.B. called him "The best infectious disease man in the world, and a wonderful and moral human being to boot."[252]

When Koenig developed lymphoma, he came to T.E.B. for help. For several months Koenig had intermittent low-grade fever and enlarged lymph nodes; T.E.B. treated him with antibiotics. The size of the lymph nodes waxed and waned, a common occurrence. At times he felt terrible, then returned to work for a few months, but his overall condition slowly worsened. T.E.B. was aware that several of his hematology colleagues criticized him for not starting chemotherapy. He was not defensive about his treatment, but he was not immune to criticism by his coworkers. Brittingham granted that he was not treating Koenig conventionally, because he would not treat himself conventionally in the same circumstances. He was afraid that conventional treatment might kill Koenig. He cited "a million bystanders" telling him how sick Dr. Koenig was and wondering what he was going to do about it. T.E.B. noted that Dr. Koenig was patient and would do whatever he advised. "But when things go wrong, there is no question who is responsible," he wrote.[253]

Dr. Sanford (Sandy) Krantz was chief of hematology at the VA Hospital during Dr. Koenig's illness. He said that Koenig's lymphoid tumor started as a giant follicular lymphoma, a type of tumor that progresses slowly for several years. In the early stages, giant follicular lymphoma is very sensitive to chemotherapy. When treated with chemotherapy, the enlarged lymph nodes virtually disappear, and the patient's quality of life improves. Accordingly, most hematologists treated giant follicular lymphoma with chemotherapy. But there was no evidence that chemotherapy extended the life of the patient. The tumor came back eventually. In Dr. Koenig's case, the tumor transformed into a reticulum cell sarcoma, for which no effective treatment was available at that time. One could argue, therefore, that there was no substantial loss by not treating Dr. Koenig's tumor with chemotherapy in its early stages, especially since treatment had many side effects, including the risk of early death.[254]

By the spring of 1972, Dr. Koenig's condition had deteriorated. His intermittent confusion, which in retrospect was due to marked elevation of calcium in the blood, perplexed T.E.B. Sometimes Koenig couldn't speak or even suck through a straw. His symptoms would come and go. T.E.B. made house calls daily.

Once he saw Dr. Koenig at midnight, because of intractable hiccups. Some of the hiccups interfered with his breathing, and his wife, Connie, had to pound on his chest. Koenig thought he was dying. He had a fecal impaction in his rectum but said he wouldn't ask anyone to break that up for him. T.E.B. pulled on a set of gloves and broke up the impaction. He had to repeat this procedure three consecutive nights. Removing the impaction seemed to eliminate the hiccups. T.E.B. said, "I got such satisfaction from helping him. We want to do something useful. It's just being of service that gives you a good feeling."[255]

As his friend was dying, T.E.B. was devastated by frustration. He said that taking care of Dr. Koenig had been the most stressful situation he had ever experienced. Koenig's weight had dropped from 175 to 83 pounds, and it appeared he would die soon. T.E.B. thought that such a terrible result was evidence that he had made a mistake, but he didn't know what the mistake was. His frustration related to taking care of a dear friend. He tortured himself, wondering if a doctor should never take care of his friends and family because it would be stressful if they did badly. T.E.B. allowed that he thought he knew as much about Koenig's problem as anyone in the world, then conceded that such thinking represented unwarranted arrogance. He finally decided that he should get his mind off himself and onto his patient. "I should just go ahead and do the rotten best I know how."[256]

Glen Koenig died in November, 1972. Dr. Alan Cohen was the resident on the service during the terminal hospitalization. He saw T.E.B. put his heart and soul into caring for his patient. Brittingham typed an eight-page, single-spaced letter to Dr. Koenig's family. He dissected his thought processes month by month during the last year. He described the agonizing frustration he had felt during Dr. Koenig's illness and acknowledged his mistakes in treatment. As a teaching effort he sent copies of the letter to all the residents who had participated in Koenig's treatment.[257]

Intense debates occurred at Vanderbilt in the early 1960's regarding the merits and possible complications of coronary arteriography. T.E.B. was skeptical about the procedure, because serious complications had occurred with early attempts in the radiology department. The cardiologists at Vanderbilt were conservative then, and the decision whether or not to introduce coronary arteriography and coronary artery bypass surgery at Vanderbilt went all the way to the Board of Trust.[258]

Dr. Mason Sones, a pediatric cardiologist at the Cleveland Clinic, had recognized the opportunity of selective coronary arteriography in 1958 when a catheter tip positioned in the aortic root inadvertently moved into the right coronary artery. He injected 30 cc of dye directly into the right coronary artery instead of the aortic root. The result showed a detailed arteriogram of the right coronary artery with a resolution never seen before. Sones helped develop a cine camera to record the images at thirty frames per second. He subsequently perfected coronary arteriography, and he and his colleagues performed the procedure on hundreds of patients. Using this technique, physicians could definitively analyze the anatomy and obstructions of the coronary arteries. Coronary arteriography spread throughout the country and became the gold standard for evaluating coronary artery disease, which heretofore could only be diagnosed by assessing the patient's symptoms and electrocardiograms.[259]

In 1966 and 1967, Dr. René Favaloro at the Cleveland Clinic worked out the details and techniques of the first successful coronary artery bypass procedures, after which coronary artery surgery became a primary treatment for coronary heart disease. Coronary artery bypass surgery was performed with grafts from saphenous veins obtained from the legs. The surgeon cut the veins into appropriate lengths, then sutured the ends of a vein into the ascending aorta and the recipient coronary artery below the obstruction. The surgeon could bypass three or more obstructed coronary artery branches with vein grafts in one operation.

The potential scope of coronary artery surgery was immense, because coronary artery disease was the leading killer in the United States and Western Europe. Virtually the entire adult population in developed countries was susceptible to the disease. Many cardiologists and internists recognized the logic and value of coronary arteriography and coronary artery bypass surgery, but a minority of more conservative physicians, including T.E.B., were unconvinced at first, and several years of controversy ensued.[260]

Following coronary bypass surgery, symptoms of angina pectoris quickly subsided. Chest pain with exertion disappeared in 90% of the patients. Virtually everyone participating in this therapy thought that surgical intervention improved the quality of the lives of their patients. However, in 1983 survival data from a large multicenter randomized trial showed no statistical difference between patients who had coronary bypass operations and those who had nonsurgical therapy.[261] There was scant data indicating that coronary bypass operations prolonged life, except in isolated instances.

Coronary artery bypass surgery first took hold outside the university setting. Its use spread rapidly through the private practice sector. It met initial resistance in many university centers, though this abated with time. At academic centers new procedures were more likely to receive scientific scrutiny before receiving widespread backing. T.E.B. knew of the effectiveness and safety of coronary artery bypass surgery, from its successful use at nearby St. Thomas Hospital in Nashville, but his conservative instincts restrained his advocacy. His controversial view on this issue did not lead to warm working relationships with his cardiology colleagues.

By the late 1970's some of T.E.B.'s former students who had left Vanderbilt for subspecialty training elsewhere returned to join the Vanderbilt faculty. Some were not as willing to accept T.E.B.'s unconventional ideas as they had when they were students. Brittingham may have felt discomfited by this. He certainly felt the insidious erosion of his status. Some of his controversial positions now seemed archaic because progress had proven them mistaken. His resistance to the growing tide of specialization, sub-specialization, and procedure-oriented internal medicine derived in part from the fact that it devalued the role of the general internist. For many years he had glamorized the appeal of just being a "country doctor." His controversial ideas played a role in his eventual decision to leave Vanderbilt.

16

Burnout at Vanderbilt

"After a lifetime of teaching doctors, he most of all wanted to be one. He wanted to be in practice." —Dr. Roger Des Prez

Why did Dr. Tom Brittingham, a legendary physician-teacher, leave Vanderbilt in 1980, move to Fort Worth, Texas, and enter private medical practice? Personal and family issues, changes at Vanderbilt University Hospital, and his perceived diminution in effectiveness coalesced into his decision.

As early as 1971, T.E.B. showed signs of burnout at Vanderbilt. He was not producing any scholarly publications and acknowledged that he may never do so again. He was taking care of a moderate number of patients, but thought he could do this more effectively in another setting. "Working in the big medical center here, I do not seem to fill any important need," he said, and "I no longer have a great deal to bring to teaching, since after eight years Vanderbilt knows what I think."[262]

David Rogers had resigned as Chairman of the Vanderbilt Department of Medicine in 1968 to become Dean at Johns Hopkins School of Medicine. He asked T.E.B. to come with him to Hopkins, but T.E.B. declined. At that time he was imbedded in his responsibilities at Vanderbilt.

When Brittingham's daughter, Susan, was applying to medical school in 1972, he wrote to her that coming to Vanderbilt might be uncomfortable for her. "I have functioned less well than formerly, and I have evoked a lot of ill will all over the medical school—principally for not agreeing with people." He added that he had not been as diplomatic as he should have been.[263] Susan was accepted at Vanderbilt the next year.

Ambivalence about his role in life had become apparent from time to time. In the summer of 1964, T.E.B. ran the medical service for six weeks during Rogers' absence from the city. His reaction to an incident which occurred during that period demonstrated how his idealism sometimes tormented him. One of the interns made a colossal error on a patient he saw in the emergency room. T.E.B. agonized that the mistake had to be pointed out in a manner that the young doctor who made it would not be destroyed, while others could learn from it. "I don't know if I will be able to do this," he wrote.[264]

After only two years at Vanderbilt, self-doubts about his role in life surfaced. He often wondered what he was doing with all his learning and why he led what he considered such a soft life. "I have the almost continuously uncomfortable feeling of not doing what I ought to be doing," he wrote. In 1974, he said he should really go out and practice medicine but that he would rather do it in some place where he could fill a big need.[265] He received an invitation to become Chief of the Medical Teaching Service at the University of North Carolina, but declined the offer.

Four years before leaving Vanderbilt, T.E.B. corresponded with Dr. Michael G. Sribnick, a former Vanderbilt house officer, and told him of his intent to enter private practice with Dr. Thomas Q. Davis in Fort Worth. Scribnick noticed that T.E.B. seemed to lack confidence about his ability to cut the mustard in private practice and acted "scared to death."[266] Scribnick was correct. In a letter to Dr. Bob Dunkerly in 1979 T.E.B. wrote, "If you still see me here in 1980, you will know that you have one good friend who is lacking in courage."[267]

The stated reason for his departure was to care for his mother, who would be nearly ninety years old when Brittingham was considering the move to Fort Worth. His mother, Lucile Brittingham, did undergo a colon resection for cancer at Vanderbilt in October, 1979. She recovered nicely, however, and outlived her son by ten years, dying at age 105.

T.E.B. with his mother, Lucile Matthews

Several other factors played a role in his departure. One was that Vanderbilt was changing in a way that Brittingham did not favor. Vanderbilt chose to participate in the Physician's Augmentation Program (PAP), a federal initiative to increase the number of physicians in the United States by 50,000. The Vanderbilt medical school class increased from 52 to 105 students. In return, Vanderbilt received hundreds of thousands of dollars a year for faculty salaries and new physical plant construction. Brittingham grudgingly recognized that Vanderbilt should take the federal money but was unhappy with the increase in class size. He believed that as class size increased, the faculty would not be able to give each student the personal attention that had been customary. He advised Dean John Chapman to "be as independent as you can be" when it came to accepting federal money.[268]

When T.E.B. joined the Department of Medicine, everyone in the department knew each other well; they were friends, and they socialized often. As the Department of Medicine grew in number, evolving from a small band of friends to a large number of strangers, it lost its inherent cohesiveness. By the late 1970s, of the original young faculty friends that Brittingham had joined in 1963—David Rogers, Grant Liddle, Robert Heyssel, David Law, John Flexner, Glen Koenig, Tom Paine, Lloyd Ramsey and Roger Des Prez—only Liddle, Flexner, and Des Prez remained.

Vanderbilt Medical School and its affiliated hospitals, as other academic medical centers, grew to gargantuan size. It ceased to be the tightly knit community that had welcomed Brittingham during the Camelot Years of the 1960's. Faculty members, consumed by the "publish or perish" ethic, grew less interested in teaching. Research and clinical revenue generated income, whereas teaching did not. There was neither time nor incentive for most senior faculty to closely mentor house staff as T.E.B. did. In this new environment patient care turned into a marketplace commodity, with patients seen as customers rather than suffering human beings.[269]

Eventually, the house staff interview process also grew to be a burden for T.E.B. He was particular who joined the Vanderbilt medical house staff and wanted to have the final say in the process. Other faculty members, seeing Brittingham so heavily involved with house staff selection, pulled back gradually and let him carry the entire load. He enjoyed interviewing, but some days he did nothing but house staff interviews. Dotsy said that, in her opinion, this was the main reason he left Vanderbilt.[270]

T.E.B. also felt he was losing some of his creativity, becoming too predictable and perhaps a caricature of the stereotype students and house staff had developed of him.[271] In a letter to Drs. Grant Liddle and John Oates he flatly stated that he had nothing worthwhile to say at a research seminar and therefore no longer belonged in academic medicine.[272]

Another trend that infuriated T.E.B. was the effort underway to make the internal medicine residency less taxing by switching from an every other night on call schedule to every third or every fourth night. Much of the excitement of medicine occurred during the evening hours. By their presence in the hospital at night, house officers could share interesting clinical experiences with their colleagues. Vanderbilt Hospital provided a snack for the house staff at 10 p.m. Over cheeseburgers and milkshakes they discussed patients admitted that evening, and afterwards could go together to listen to an unusual heart murmur or feel an abdominal mass.

T.E.B. vigorously resisted any change to the every-other-night resident call, but some of his faculty colleagues, including Dr. Gottlieb Friesinger,

Director of the Division of Cardiology, favored the proposed change. They argued that other academic medical centers were out-competing Vanderbilt in attracting good residents because of their more lenient call schedule, and they also cited the issues of resident fatigue and patient safety.[273] These arguments held no sway with T.E.B.. Supporting his assertions, surveys of residents considered the lack of faculty supervision and the handoff problems (handing off the patients of the resident leaving for the night to another on call for that night) as more serious causes of physician error than fatigue.[274]

T.E.B., who was aware of most of the house staff mistakes, did not think many of them resulted from fatigue. He wrote to Dr. Grant Liddle, Chairman of the Department of Medicine, that the proposed change would initiate an irreversible process of decline. "I suspect that if I come back to Vanderbilt five years after this change is initiated, I will find less good house staff and students than we have right now—they're pretty good right now." He stated that he would not recommend Vanderbilt for someone who was slow or sick or not interested in trying to achieve excellence. In T.E.B.'s opinion, the proper way to take care of patients was to be on call for them every night, and he planned to put this into practice for himself.[275]

Photo by Dr. Eric Dyer

Dr. Grant Liddle

T.E.B. also opposed the plans to construct a new and bigger hospital, promoted by the Vice-chancellor for Medical Affairs, Vernon Wilson, but this was another battle he knew he would lose. With the physical plant, as with class size, Brittingham did not believe that bigger was better. He wrote to a member of the Vanderbilt Board of Trust, "I believe that too much of our GNP is going into health care right now, that building a new 60 million dollar hospital will only accentuate this problem. Of course this view is not likely to be widely shared by other doctors."[276] As Wilson forged on with plans for the new hospital, Brittingham met with every trustee he knew——and gave each of them a copy of E. F. Schumacher's book, *Small is Beautiful*.[277]

T.E.B. liked the layout of the familiar old hospital, which mimicked Hopkins and Barnes in placing his and many other faculty offices close to the sites of physician training. In the smaller, old hospital, casual, water cooler conversation and consultations occurred regularly. The new hospital was remote from faculty offices and did not encourage random encounters or residents dropping in to ask questions.

One of the biggest personal issues for T.E.B. was his perception that his influence at Vanderbilt had lessened. The role of the general internist had decreased, and specialists had gained control. CT and MRI scans, endoscopies, and arteriograms often solved diagnostic problems readily, which resulted in a lesser emphasis on the fundamental skills of medical history-taking and physical exam.

A perception of failing health may have motivated him, or at least encouraged him to reorder his priorities. In February, 1979, while taking a five-mile walk with Dotsy and his daughter Margaret in subzero weather to Picnic Point at Lake Mendota in Madison, Wisconsin, T.E.B. developed pain in his hip. Later he would attribute this pain, in retrospect, to metastatic renal cell cancer. He returned to the hotel room to rest and watch a hockey game, because he didn't feel like walking further.

Two letters T.E.B. received at this time are of particular interest. David Rogers wrote how much he respected T.E.B.'s decision that he wanted to do something different, but he hoped that T.E.B. had also weighed his "remarkable clout on medical students and house staff" against his skills as a practitioner. As before, Rogers was concerned that

Brittingham's life should be as much fun as it should be. Rogers wrote, "If you want to change, go to it, I'm with you all the way."

Pathologist Bob Collins, who had argued with T.E.B. incessantly, if not successfully, that Hodgkin's Disease was indeed a neoplasm, wrote, "Certainly there will always be misconceptions about medicine needing the fresh approach you have exemplified." Collins stated his appreciation of the positive influence T.E.B. had on himself and Vanderbilt, and reminded him that his many friends in Pathology would remember the excitement and honesty T.E.B. had instilled in their practice.[278]

At T.E.B.'s last grand rounds at Vanderbilt, he presented the case of a 51-year-old black lawyer and former football coach, who arrived at the Vanderbilt emergency room on a Sunday afternoon in shock, with cold, clammy skin, pulse rate of 150, and no detectable blood pressure. He had gone to the home of the Fisk University president to congratulate him on Fisk's undefeated season, had a drink or two there, then had sudden onset of severe lower right abdominal pain and fainted.

An emergency upper endoscopy showed erosions in the stomach and duodenum, consistent with recent alcohol ingestion, but no evidence of severe or acute bleeding. He received ten liters of intravenous saline and 4 units of blood and improved overnight, in fact he sat up to eat breakfast. But later in the morning he had recurrence of severe right lower abdominal pain and went into shock again.

That evening an arteriogram showed a ruptured abdominal aortic aneurysm. Surgery to repair the aneurysm began at 10 p.m. and lasted eight hours, during which he received 40 units of blood. The surgeons constructed a dacron graft of the lower aorta, but the patient died later that day.

The patient's obituary in the Tennessean said he died at Vanderbilt Hospital of complications arising from an aneurysm of the aorta. T.E.B.'s final note on his typed lecture notes said, "Principal complication may have been our failure to make the correct diagnosis promptly." T.E.B. must have thought that an arteriogram could be a wonderful, lifesaving diagnostic procedure, but only if the doctor thought to order it.

T.E.B.'s decision to depart from Vanderbilt unveiled his disquiet about some personal and professional concerns—his mother's health, the growth of the faculty in the Department of Medicine, the decline in influence of general internal medicine, the new hospital, the shift away from every-other-night call, and his controversies with certain faculty members. But it ignored other indispensable elements in his relationship with Vanderbilt—his teaching, his influence on students and residents, and his care for the well-being of patients. In a sense, this was because by the time he departed from Vanderbilt, the inexorable tide of specialization, technological procedures, and commercialization of medicine had already rolled past.

What was the effect of David Rogers on Brittingham's career at Vanderbilt? It came just at the right time, when his post at St. Louis City Hospital had no future and was not supported adequately by Washington University. T.E.B.'s goal was to pursue doctoring to the best of his ability, without distraction. The position at Vanderbilt with David Rogers was not only ideal but also career-galvanizing for T.E.B.. His brilliance in teaching was able to achieve its full potential there.

At the height of his influence he was able to doctor—to teach doctoring and to practice it. But by the time he left Vanderbilt, new technology had overshadowed doctoring, and T.E.B.'s methods often seemed dated and out of place. His unique style of doctoring, which once had sustained him, had virtually disappeared and was alive only in the minds of his disciples. Physicians' house calls were replaced competently by home health nursing and hospice care. T.E.B.'s career followed a patient-oriented, late-nineteenth century, Oslerian model. He had won no research prizes nor made new medical discoveries.

When Brittingham announced to Grant Liddle that he was leaving the Vanderbilt faculty, Liddle begged him not to go. Liddle told T.E.B. he could do anything he wanted to do in the department if he stayed, but Brittingham had made up his mind and departed from Nashville January 1, 1980.

17

Practice in Fort Worth

Medicine was everything to him—his vocation and his avocation—and he never tired of it.—Dorothy Mott Brittingham

In 1979 T.E.B. called his old friend from Harvard Medical School, Dr. Robb Rutledge, a surgeon in Fort Worth, and informed him that he was planning to move to Fort Worth to practice medicine. He told Rutledge, "I've never been a real doctor in my life, and I want to see what it's like."[279] So at age 57, Brittingham joined Internal Medicine Associates of Fort Worth, then considered the best internal medicine practice in the city.[280] Although his wife Dotsy was at first reluctant to move—she was settled and content in Nashville, where their youngest child Sally was still in high school—they moved into the third floor of the house of Lucile Brittingham, T.E.B.'s mother, at 3 Westover Road in Fort Worth. Most days T.E.B. ate a sack lunch, brought from home, but Dotsy frequently brought a picnic lunch, and the two of them ate together, standing beside his Thunderbird, using its hood as a table.[281]

T.E.B. loved practice, though he was probably ill and certainly in increasing pain the entire time. In contrast to Vanderbilt, where he often took controversial positions, in Fort Worth his approach to patients and his medical colleagues was conventional. He was exceptional, however, in maintaining total availability to his patients. He did not sign out to his partners on weekends and almost never left the city. Dr. Rutledge said that much of T.E.B.'s practice consisted of older people, whom he treated with unusual warmth and kindness. He made house calls every Saturday afternoon and frequently during the week. When he rounded in the hospital, he carried his black bag and his old typewriter, the only doctor to do so. He was a favorite with the nurses, who liked him so much that they even learned to read his writing. He felt like a real doctor.

The other physicians in the group included Thomas Q. Davis, Ed Nelson, and Ed Forshay. Davis and Nelson practiced general internal medicine, while Forshay was an endocrinologist. Davis had completed his internship and two years of residency at Vanderbilt, then had served as Chief Resident at the Nashville VA Hospital. In fact, Brittingham had selected Davis for internship in 1969. He had acquired the nickname of "Mad Dog," because of his aggressive style of teaching. One of his duties at the VA Hospital was to interpret all the electrocardiograms. If he found an alarming EKG, he did not simply submit a report, but went directly to the ward and met with the intern or resident caring for the patient, making sure that the house officer had appreciated the EKG abnormality and had taken appropriate action. Brittingham had admired Davis's active style, and when he decided to move to Fort Worth, he contacted Davis about joining his practice. Davis said he was stunned when Brittingham called. After T.E.B. joined the practice, he often dropped into Davis's office to ask, "Mad Dog, how do you do this?" Davis was overwhelmed. "Here is the guy I worshiped asking me how to do something."[282]

On one occasion, one of Davis's patients was dead on arrival at the Harris Hospital Emergency Room (now Harris Methodist Hospital of Fort Worth). Davis had pulled on his clothes so fast that he wore mismatched shoes, one brown and one black and of different styles. Dotsy told T.E.B. that could never happen to him. All his shoes were the same, some were just more worn out than others.[283]

Ed Nelson had worked with the Indian Health Service in Oklahoma, where he enjoyed participating in surgery and developed an interest in obstetrics. Later, he started an emergency room practice in Fort Worth. In 1976 Brittingham interviewed him for a residency in medicine at Vanderbilt. Nelson described the interview as pleasant, but thought that Brittingham questioned his commitment to internal medicine because of his surgical and obstetric interests. He did his internal medicine training at Parkland Hospital in Dallas. Nelson described Brittingham as a "delightful guy, cordial and friendly," but added that T.E.B. was passive in his engagement with his colleagues. Though their offices were only a few feet apart, Nelson related that he and T.E.B. rarely interacted. "I had lots of respect for him, I had the notion this was a great man," Nelson told me. Nelson said T.E.B. worked his ass off and that he usually stayed at the office later than the other associates.[284]

T.E.B. jumped into the rhythm of private practice. One evening he came home from the office and was so sick that he skipped dinner with guests and went to bed. But it was his night on call for the group, and he received an emergency call at 11:30 p.m. It concerned a partner's patient who had had routine surgery that afternoon and later had stopped breathing. He got up and went to the Intensive Care Unit. He wrote his daughter Sally (one of 82 letters to her within six months), "To my great surprise my body, sick as it was, could handle the stress without any trouble, so I think I may be able to handle the hours here. It's actually exciting to do something different from what you have done for years."[285]

Davis said that the patients who gravitated to Brittingham were those who had seen every other physician in North and West Texas and whose diagnoses were still elusive. He told me that he felt blessed to have a personal relationship with T.E.B., at least as far as it was possible to have a truly personal relationship with this extremely private man. As compassionate and caring as he was, Brittingham had an inner toughness that he had inherited from his family. While T.E.B. seemed happy to be out of academia, Davis said, "I think the academic setting, particularly Vanderbilt, was where he was at his best."[286]

Brittingham took call for his partners every fourth weekend, but he didn't allow them to take call for him—not because he didn't trust their competence, but because he believed in always being available to his patients. After a few months in practice, T.E.B. wrote, "In Nashville I never charged patients anything, here I am charging them a lot, that changes the relationship quite a bit, they are prone to see you as a money-hungry doctor whatever feelings you may have on the inside."[287]

"The chief complaint of most of my patients is that they couldn't get in to see Dr. Davis or Dr. Nelson, so they had to come see me," T.E.B. said. "In Nashville, I had a lot of prestige (I think), and that tends to make you look good even if you aren't doing a good job. Here I have zero prestige, you have to be able to cut it completely on your own."[288]

Although he got along with his colleagues, he did things his own way, which occasionally appeared out of step. One of his patients, a frail, eighty-year-old man, had an episode of chest pain at home. Rather than admit the patient to a coronary care unit, which was customary at that time, T.E.B. took an EKG machine to the patient's house, confirmed the diagnosis of acute myocardial infarction (heart attack), had oxygen delivered to the home, and made house calls twice daily.

T.E.B. saw fewer patients than his partners, primarily because he spent unrestricted time with each patient. He never used patient questionnaires or other shortcuts to promote office efficiency. He showed little interest in monetary production or other financial aspects of private practice.

Because Brittingham's practice style was so different, after his first year Internal Medicine Associates amicably separated the business interests of their practices from T.E.B. He continued to rent office space from them, but maintained a solo practice. He wrote, "I would have made a mess of office management if left alone."[289]

T.E.B.'s evaluation of practice did fluctuate. At one point he had 17 patients in the hospital. He frequently was awakened at night to see patients in the hospital or emergency room. One Sunday morning he interrupted his hospital rounds to go to church, then returned to finish rounding in the afternoon. "I went to church not to do God's will, but strictly because I had to for myself," he said.[290]

T.E.B. wrote, "I don't get any chance to either read or think, and pretty soon I believe I will go to pot."[291] A few months later he added, "As far as I can tell the hard work doesn't hurt me ... perhaps because I think I'm doing what I'm supposed to be doing, who knows? It doesn't seem very intellectual ... it's just my little corner of life, and I don't intend to half-ass it."[292]

David Rogers wrote to T.E.B. about his concern regarding T.E.B.'s attitude and approach to practice. He worried about the personal costs for both T.E.B. and his family stemming from the way T.E.B. had decided to play his role. Rogers felt his friend was working at doctoring to the exclusion of all other parts of his life and didn't think that was smart. He feared T.E.B. would burn out and that even his doctoring would suffer if he didn't pace himself better. Imagining himself in the role of T.E.B.'s patient, Rogers said he wouldn't give a damn about his doctor *always* being available. It wouldn't trouble him a bit to see an assigned colleague because T.E.B. had taken a weekend off to climb a mountain at Lambshead Ranch, go to the opera, or take a trip. Rogers insisted that he'd feel *more comfortable* putting demands on his physician if the physician had been strong enough to plan some time for himself as well as his patient. Rogers ended the letter as follows: "Enuf—you're a wonderful guy and our long and close relationship is very special ... thus I felt it reasonably safe to gently kick your butt."[293]

After eighteen months in practice, T.E.B. felt less stressed and called his life "a joy." His letters to his children were more contemplative, almost ruminative. He spoke more often of religious thoughts and of the importance of good health. "I never think of death as a tragedy of any kind," he wrote to his son Tommy. A few months later he added, "I think we must feel guided by some Higher Being."[294]

He said that he worked hard, not to make a lot of money or to gain a big reputation, but just to do the job as best as he could. "I'm going to try until I burn out or rust out or drop dead or whatever." He acknowledged that he did burn out at Vanderbilt, but had no desire to return to teaching.[295]

In a letter to his son Tommy, then attending University of Tennessee at Knoxville, he wrote, "I keep searching for meanings in life—I would like to think that our first loyalty is to our fellow man, period, and not to our nuclear family ... You and your siblings still have a father, he is just in a different location than where he used to be."[296] He had missed Thanksgiving dinner in Nashville the previous year because he was on call for his group.

After two years in practice, T.E.B. boasted that he had not been out of the county for the past 730 days.[297] He said he was the only Fort Worth internist practicing solo, but he recommended it heartily, and after three years he prided himself on practicing all alone, on call 24 hours daily, 365 days yearly: "It does not bother me a bit ... one gets enough sleep, best way to practice medicine ... I think it is possible we shall learn the Great Lesson of Servanthood."[298]

Despite these varied reports, Dotsy told me it was her opinion that T.E.B. was happy in practice and with life in Fort Worth.[299]

Practicing general internal medicine did have its hassles. In a letter to Dr. Brittingham the Chairman of the Medical Staff Quality Assurance Committee at Harris Hospital suggested that a particular patient was not an appropriate person to be admitted to Harris Hospital. If Dr. Brittingham wanted to admit her in the future, he would need pre-admission approval from a member of the Internal Medicine Quality Assurance Committee. T.E.B.—the old T.E.B. from Vanderbilt—was infuriated and protested vigorously:

> Dear Dr. X,
>
> Today I received your Quality Assurance Committee's letter about Carla Johnson ... If you mean Harris does not like to take care of patients like her, I don't blame them. I don't either, as can be seen in my two admission notes (2-28-84 and 5-6-84). I personally consider her to be very demanding, often hostile, litigious, unrewarding financially, difficult to understand medically, complicated, very possibly having severe psychiatric disease, a doctor shopper of high degree, perfectly willing to take advantage of other people, and sick. Does your Committee mean that Harris should pick and choose whom it cares for? That I should pick and choose whom I care for? Perhaps sticking to white residents of Westover Hills who have well integrated personalities and are not very sick? I don't believe we are permitted to do that, I don't believe you do either. I feel ashamed for making negative observations about Carla Johnson, but they do seem quite blatantly correct.
>
> Do you mean she should not be getting pain medications, that giving pain meds to this woman activates the Quality Control apparatus of Harris? This is a reasonable point of view. However, she had a lumbar laminectomy and fusion in 1966, similar surgery in 1973, similar surgery again in 1976, and similar surgery again in 1979. When seen at Harris by Dr. Levy in June '83 a myelogram showed severe scarring at multiple levels ... and a possible disc bulge at L3-4. Dr. Bechtel considered her inoperable. This is an excellent background for pain,

sounds like she has adhesive arachnoiditis to me. In Oct. '80 she had upper tibial osteotomy on left knee, in July '81 she had left total knee replacement, in Nov. '82 she had revision of left total knee and then post-op required knee manipulation under anesthesia....What to do when someone with this background complains of pain??? My inclination is to give analgesic, I think it is a very reasonable point of view ...

Do you mean I should have sent her to a psychiatrist? She has been seeing a psychiatrist, Al Marshall. She had severe depression in Nov. '63 with a long hospitalization, and I think got electroshock therapy then...From 7-25 to 8-19-83 she was hospitalized at St. Joseph Hospital in the Pain Program. After a month the conclusion was as follows. 'The patient was not asked to return as an outpatient to the Pain Program. It was the general impression in the Pain Program that there had been no improvement and no success with this patient. It was our anticipation that she would be looking for another physician to give her pain medication and to see about some type of surgery.' Guess what happened—I got her back. I have at least kept her from surgery. This woman has seen psychiatrists, probably lots of them. They have not helped her. I hope her present stay in St. Joe Psych helps her, but I rather doubt that it will. If you or your committee can tell me which psychiatrist or which Pain Program can fix this woman, I shall be happen to listen.

I think a reasonable reason for activating Quality Assurance is a sloppy and incomplete workup, but I doubt that is what brought me your letter. If you read my 2-28-84 workup, I reviewed in detail all the past Harris admissions that I could find from 1959 on. I found 16 previous Harris admissions, and Medical Records probably complained about finding those for me, they sometimes do complain. I am interested that she has had 34 admissions to Harris since 1976 and hope that you will send me a list of those

34 admissions because I sure looked for them. It sounds like perhaps Quality Assurance should look at Medical Records for a moment. Incidentally one of those old admissions ... was by you, so you are no more than 1 admission behind me. And incidentally the reason I read all those confounded old records was in an effort to better understand Carla Johnson. If the nature of her problems is obvious to the neurologist and orthopedist and other doctors who saw her and to you and to the Harris Internal Medicine Quality Assurance Committee, great. I always knew there were a large number of doctors a lot smarter than I am.

Your letter notes that there must be pre-admission approval from a member of Harris Internal Medicine Quality Assurance Committee before Carla Johnson is admitted again. Was the problem that her clinical situation was an inadequate reason for admission????? I have admitted her to Harris just twice in my life. Once was 5-6-84. I visited her at home 5-5, she looked terrible, I told her I thought she needed to be in a nursing home, which she resented. I thought her problem was toxic encephalopathy from excess drugs given her by the neurologist who had recently had her in Huguley Hosp. and just released her on 4-25-84. I pleaded with her to let him deal with the problem, it was really his. So what happens after the Saturday house call? Her Harris CareLine gets activated and at 1:05 a.m. on Sunday 5-6-84 she is dropped in the Harris ER and at 1:30 a.m. Sunday 5-6-84 the Harris ER calls to tell me she is all mine and she is going to have to be admitted. At 5 a.m. Sunday, 5-6-84, I was called and told that she was very hostile and demanding and could she please be given some Demerol. Harris Hospital, I love it.

My only other Harris admission was 2-28-84, when again her Harris CareLine was activated, firefighters broke into her home and brought her to ER, I was called at midnight that she had arrived and then at 1:10 a.m. to be told that she needed admission (in my opinion she did). She had been found lying face down on the floor of her home covered with vomitus. There was vomitus in her mouth and all over her hair, she may well have aspirated vomitus. In the ER she had poor color and was hypothermic with rectal temp of 94.3. Do you really think this admission was unwarranted? Did you or your committee read her charts carefully before you wrote that letter??? Note that both these admissions were brought about by the Harris CareLine, better get rid of it for Carla Johnson, she is not deserving of it, it's more for people like you and me ... Both admissions were really dumped on me by the Harris ER, why don't you just have Quality Assurance tell the ER to be a little sterner about disposition of problems from the ER?

I would like to suggest to you that thoughtful physicians can look at the same patient in different ways. Who knows what the truth is, but there is room for honest differences of opinion in medicine, we all need not think like your Internal Medicine Quality Assurance Committee. Or must we all think as it does??

At your committee's leisure I will appreciate knowing exactly what generated your letter to me. You have left me wondering how useful committees are.

Tom Brittingham[300]

*****

On April 22, 1986, Dr. Robb Rutledge was seeing patients in his office. He took a telephone call from Dr. Tom Brittingham. In a calm, almost matter-of-fact voice, T.E.B. said, "Robb, I have metastatic cancer of the kidney. I'm not going to live long, and I want you to sign my death certificate."[301]

T.E.B.

Earlier that day, after years of pain in his hip which resulted in a progressively severe limp, and at the recommendation of his son-in-law, Dr. Clark Gregg, T.E.B. went alone to the radiology department at Harris Hospital and arranged with Dr. John S. Alexander, the radiologist assigned to computed tomography (CT scanning) that week, to undergo a scan of his left hip. The scan revealed a kidney cancer with metastases to the liver and bones. He accepted the findings soberly and calmly.[302]

For seven years T.E.B. had noted pelvic pain. In 1980, while carrying a heavy bag through an airport, the pain suddenly became much worse. From that time on, he was unable to run. The pain slowly progressed. In early 1986, he developed a persistent cough, fatigue, weight loss, and afternoon fevers. He wrote to his former resident Charles Bryan, who had recently published a paper describing three patients with fever of unknown origin due to cytomegalovirus, and requested a reprint of the article.[303] He thought he might have something like cytomegalovirus and that it would go away.

T.E.B. drove to his office after the CT scan, arriving just as Tom Davis pulled into the parking lot. Davis, who had noticed what he called a "gimpy limp" since T.E.B. joined the practice, thought the limp appeared worse. He asked T.E.B. when he was going to get someone to look at his hip. Brittingham replied, "Great man, I just did. Come to my office. I'd like you to sit with me when I tell Dotsy about it." Before Dotsy arrived, Brittingham informed Davis he had cancer, then Davis sat "dumbstruck" as T.E.B. told Dotsy about his illness.[304]

Brittingham saw patients the remainder of the week then closed his office. He never saw another patient nor read another medical journal. He moved his patient records to his home, completed his Harris Hospital records, and resigned from the hospital staff. He arranged for other physicians to care for his hospitalized patients, and, over the next several days, wrote personal letters to each of his patients. He did not leave the house again. He accepted his illness without fear.

He made several telephone calls to urologists he trusted, and on the basis of these and his own inclinations, decided against any attempt at treatment. He consulted no oncologist or other physician. He did not put himself under a doctor's care, though during his decline Robb Rutledge visited frequently and assisted as he could.

His Hotchkiss roommate, Don Durgin, called to express concern and said, "Isn't there an expert you could see?" T.E.B. answered, with a chuckle, "Don, no offense, but I am an expert."

During the months preceding his death, Brittingham wrote dozens of letters to many of his former residents and friends, expressing his love to them and how fortunate he had been to practice medicine. Among many others, Drs. Taylor Wray, Dick Dixon, Bob Dunkerly, and Grace and Tom Paine received such letters. In May he watched the NBA playoff games and the World Cup on TV.

Brittingham wanted to spare his family the experience of a prolonged death. He deliberately began to starve himself. His will overcame hunger but could not overcome thirst. He could walk until June 6, when he sustained a pathologic fracture of his hip. (Pathologic fracture means a fracture which occurred spontaneously, without trauma.) After that he was bedridden.

In early June Dr. Hugh Chaplin and his wife came from St. Louis to visit. Hugh asked if T.E.B. had thought much of his death. T.E.B. said he had thought a great deal about it and concluded that he could know absolutely nothing—but that he was convinced that he would be treated appropriately—and with love.[305]

All his children visited, as did David Rogers and his second wife Bobbie, and Drs. John Flexner, Robert Collins, and Roger Des Prez from Vanderbilt. He referred to these visits as "a lot of fun."[306] Among the books he read in his final weeks was Alexander Solzhenitsyn's *Cancer Ward*. He died on July 27, 1986. Throughout his illness Dotsy attended him day and night, never left the house, and rarely left his side.

Brittingham was buried in a small cemetery on Lambshead Ranch. Ten years later his mother, Lucile, was buried next to him.

T.E.B.

Photo by Laura Wilson
T.E.B.'s grave in family cemetery at Lambshead Ranch

18

Influence and Legacy

"T.E.B. would have loved CT and MRI scans"—Dr. Charles Mayes

INFLUENCES ON INDIVIDUAL DOCTORS

Medical students and house officers encountered T.E.B. in a learning environment, during an impressionable period of their lives, when they were still wet behind the ears. Did his influence persist later in their careers? How did it affect their attitudes toward patients and inform their practice of medicine? To what extent have T.E.B.'s principles survived the changes in medicine over the intervening decades?

After Dr. John Sergent graduated from Vanderbilt Medical School, he was an intern on the Osler Service at Johns Hopkins Hospital. One night he admitted a sick woman. As he was writing up her chart about 3 a.m., he realized he had forgotten to do the pelvic exam, a requirement on all female patients. Though her illness did not involve the genital system, he started to write, "Pelvic exam deferred until morning." Then he thought about Dr. Brittingham: *I knew that if he ever became a visiting professor at Hopkins, he would see that chart, and he would be ashamed of me.* So with help from the nurse, he got the sleepy woman out of bed at 4 a.m. and did a pelvic exam—the results of which were perfectly normal.[307]

Another former medical student who studied under T.E.B. was working at a Veterans Administration Hospital. The last patient of the day was an elderly man complaining of "spells." The young doctor thought the man was just hoodwinking him and was about to send him away when the man suddenly lost consciousness, jerked his neck backward, and fell to the floor in a generalized convulsive seizure. Afterward, the doctor said he had the weird feeling that Dr. Brittingham had put that patient there as a lesson to him.

Sometimes T.E.B.'s influence took longer to surface. During his first years in practice, Dr. Karl Vandevender said that patients who didn't have a personal physician often came to Nashville hospital emergency rooms. If patients required hospitalization, the physicians who were new to private practice rotated the responsibility to assume their care. It was long before the days of hospitalists. Vandevender related that an emergency room would call to inform him of a patient who would be admitted under his care. The emergency room physician might say that the patient was stable and could be seen in the morning. But Vandevender had a premonition that some night Dr. Brittingham would disguise himself as one of these patients to find out if he would get out of bed and come to the emergency room. So Vandevender always got up and saw those patients, out of fear that T.E.B. would be checking whether he was caring for patients the way he had been taught.[308]

Others experienced T.E.B.'s influence throughout their careers. Dr. Jimmy Sullivan had been the chief medical resident at Nashville General Hospital, then had practiced internal medicine and endocrinology in Nashville for years before joining the faculty at Meharry Medical College. "I decided to become a doctor while still residing in my mother's uterus, at about eight months of her gestation. When I was a little boy, my father, a family practitioner in South Carolina, took me on his house calls," he said. "We were taught medicine by the great generation of clinical teachers. T.E.B. made being a doctor sound like fun, and it was fun. I learned from him that practice was a pure pleasure, not a chore or a way to make money. I learned that my only obligation was to the patient."[309]

Dr. Harrison Shull Jr., who had been a medical student and house officer at Vanderbilt and then practiced gastroenterology in Nashville, experienced three distinct stages of his relationship with T.E.B.:

1. The fear and awe stage, when he felt Brittingham knew everything and could do no wrong.
2. The disappointment stage, when he discovered that Brittingham did not know everything and made mistakes.

3. The appreciation stage, when he understood that Brittingham had made him think, challenged him to support his beliefs, and taught him to be critical.

During T.E.B.'s era many other hospitals had rigorous training programs. One of them was Duke University, where Dr. William Stead influenced his residents using a modus operandi different from T.E.B. Stead listened to a resident present a patient admitted the previous evening and asked, "What did the spinal fluid show?" The exhausted resident replied that he'd been up most of the night with other patients needing equal attention. "Doctor," Stead lectured, "You're telling me that life is hard. I already know that. I want to know what the spinal fluid showed."[310]

"There was a uniqueness to that era of clinical teachers," said Dr. David Robertson, a former Vanderbilt Medical Student and longtime director of the Vanderbilt Clinical Research Center "But if T.E.B. practiced in the 21st century, he would have displayed his unique gifts as he did in the 1960's and 1970's. He would have adapted to new technology because it helped patients. Before I left Vanderbilt for my internship at Hopkins, T.E.B. told me I should work faster—to speed up a bit."[311]

Dr. John Dixon was a resident in medicine and was sitting in medical grand rounds next to his intern, Sam Ashby, when Ashby suddenly had a generalized convulsion. Ashby was carried out into the hall, then wheeled on a stretcher to the Medical Intensive Care Unit. After several minutes, he gradually regained consciousness. T.E.B. patted Ashby on the shoulder and reassured him, "Don't worry, Sam, you're going to be just fine." Dixon said, "I use that technique. I pat my patients on the shoulder and tell them they're going to be just fine, just like T.E.B. did." At a later date, when he was practicing cardiology, Dixon was sitting down to Thanksgiving dinner with his wife, children, parents, and extended family. "I got an emergency call and had to leave," he said. "I remember thinking it was OK, because Dr. Brittingham would have told me to go, and somehow I felt he knew it."

One Sunday afternoon, Dixon suspected his son might have appendicitis and brought him to the Vanderbilt emergency room. A nurse practitioner saw him and ordered a CT scan. When the surgeon came in, he told Dixon his son had appendicitis and needed immediate surgery. He had not yet seen the CT scan nor met the patient. "I knew that CT scan provided an accurate diagnosis, probably more accurate than the physical exams we relied on in our day," Dixon told me. "It's hard for us old doctors to admit that, because to me the fun of medicine was the mystery of it. There was no mystery that Sunday, and my son had the prompt surgery he needed."[312]

Forty years ago, when T.E.B. taught a generation of doctors how to practice medicine, he was a shining star, remembered and revered by those he mentored. But medicine has changed. What are his enduring influences? When I posed this question to Dixon, he replied, "T.E.B.'s enduring principle is that you are a servant. You're not there to be served."[313]

Dr. Eric Dyer, a former medical student and house officer under Brittingham, was practicing pulmonology when he consulted on a man who had been hospitalized five times with pneumonia. Each time the pneumonia had resolved promptly, but nobody found evidence of an infection. His doctors couldn't understand the cause of the repeated episodes of pneumonia. In the fashion he had learned from Brittingham, Dyer sat down at the bedside and talked with the patient. He just listened—no laboratory tests or X-rays. The patient told Dyer that he had suffered from recurrent bouts of prostatitis, which always responded to an antibiotic named Macrodantin. None of the doctors who had treated the patient for pneumonia had learned this history. Macrodantin can cause a hypersensitivity reaction in the lung that resembles pneumonia but which resolves when the drug stops. Dyer was doctoring the way his mentor, T.E.B., had taught him.

Dr. Clif Cleaveland, one of T.E.B.'s former chief residents at Vanderbilt, said that the doctor he tried to be in his thirty-five years of practice was based on T.E.B.'s lessons and inspiration. "I recall the monthly cartoon *Watchbird*. A watchbird sat on your shoulder and prevented you from doing anything wrong," said Cleaveland. "I viewed Dr. Brittingham as a watchbird on my shoulder. We named our son after him: Thomas Britt Cleaveland."[314]

ALAN L. GRABER, MD

A DAY WITH A HOSPITALIST

Cleaveland designated his former medical partner, Dr. David Dodson, as the doctor most typical of practicing medicine with Brittingham's principles. Now a hospitalist, Dodson had been both a medical student and resident under T.E.B., but his own instincts, like those of others in this book, may have preceded contact with T.E.B. Dodson's father, a family physician, exposed his son to doctoring at an early age. Hospitalists are internists who are salaried employees of a hospital. They have skills and experience in the treatment of acute illnesses and critically-ill patients. Hospitalists work in defined shifts and hand over care of their patients to a colleague when they have completed their shift. They do not see patients in an office or clinic. When a hospitalist discharges a patient from the hospital, he refers the patient back to his primary care physician or arranges for other outpatient follow-up care.   Since I was not familiar with the details of a hospitalist's practice, I shadowed Dodson for a day as he made his rounds at Memorial Hospital in Chattanooga. Dodson spent at least twenty minutes in each patient's room. Starting at the foot of each patient's bed, he examined the feet, legs, abdomen, heart, and lungs. Sometimes he needed help from a family member or nurse to position a weak or unconscious patient for an adequate chest exam. He looked in each patient's mouth with a tiny LED flashlight. "I see as many patients as my colleagues," Dodson told me. "But I'm in the hospital two to three hours longer than they. Doctoring is about listening and paying attention to the patient." During the day I shadowed him, Dodson discussed the knowledge explosion of the past thirty-five years. "We can do more now than we could in Brittingham's day. I appreciate the technology that's available. But it's no longer possible to get your arms around all twenty sub-specialties of internal medicine. That's why so many of our colleagues have entered sub-specialties."[315]

When asked about the sub-specialists' relative lack of interest in the whole patient compared to the technical procedures they perform, Dodson replied, "I actually feel sorry for them. They're missing out on the doctoring part of medicine." Dodson had been so busy doctoring he hadn't taken the time to shop for a new car. His Honda Accord was twenty-six years old in 2016, even older than T.E.B.'s Thunderbird.

T.E.B.

The hospitalist model has resulted in a reduction in length of hospital stays, in rates of readmission, and in costs of hospitalization. But the impact of hospitalists on overall health, total healthcare costs, and the well-being of patients and physicians is not yet known. A patient may feel disconcerted when his personal doctor, the physician who knows him best, does not see him at his moment of greatest need. On the other hand, physicians who never see outpatients may be handicapped by not understanding patients' lives outside the hospital. In an article in the New England Journal of Medicine, Gunderman asks a fundamental question about the role of the hospitalist model in medical care: "Is the essence of medicine an institution, or a relationship between two human beings?"[316]

SPECIALIZATION

The growth of specialization has highlighted the long-standing tension between the holistic and the focused approach to patient care. Since the mid-nineteenth century, medical leaders have warned against adopting too narrow a view of the patient.[317] But today most physicians find specialty practice more fulfilling intellectually and more rewarding financially than general medicine.

Specialization in medicine can lead to gaps in care. An ophthalmologist at Vanderbilt transferred a patient with tetanus to the medical service. The patient had a serious eye infection and had undergone extensive eye surgery. The ophthalmologist had neglected to arrange for tetanus immunization. Although T.E.B. had been a specialist himself, he voiced criticism. He didn't disapprove of becoming an expert in a specialty, but he disapproved of specialists who neglected basic doctoring. He believed that all doctors should ensure that their patients were immunized against tetanus.

Dr. William Schaffner, who was chief resident in medicine under Brittingham in 1968-69 and currently is a Professor of Medicine and past Chairman of the Department of Preventive Medicine at Vanderbilt, believes that T.E.B. exemplified "The eternal verities of patient care." When Schaffner was a fellow in the specialty of infectious diseases, colleagues often said he had deeper relationships with patients' families than was expected for consultants. "That was T.E.B.'s influence on me," he said. "Some specialists consider themselves more technicians than physicians. They're dedicated to providing highly-skilled, effective care, but only within the confines of their specialty. Many think of themselves as practicing medicine just within a certain domain. And they stay securely within their domain. "We're not just grumpy old men," Schaffner explained to me. "We're concerned about our profession. Some of us thought it was a calling. Now we see many who look at it as a job. There's a hell of a difference. Medical students come with high values and high expectations. If we socialize them into thinking it's a job, that's our fault. That's because we're not teaching in the Brittingham mold."[318]

THE GENERATION GAP—CHANGING ATTITUDES TOWARD WORK AND LIFE

A number of physicians worry that *professionalism* has decreased in recent years, possibly due to the corporate nature of medicine. *Professionalism* results from a personal transformation that occurs during the early years of medical training. As part of providing care and witnessing others' suffering, doctors accept that they now interact with society in a new and different manner. As a result, physicians enter into an implicit contract to put the care of patients first.[319] In my opinion, this transformation makes it impossible to believe being a physician is *just a job*.

"Brittingham exemplified professionalism—service that transcends self-interest," wrote Dr. Charles Bryan. "When it came to medicine or to his patients, Dr. Brittingham made few if any compromises."[320]

Beginning in the 1970's *lifestyle* considerations challenged the claim of Dr. William Osler from the 1890's that the *master word in medicine was work*. Traditional attitudes toward work among young physicians began to change. One reason was the entry of larger numbers of women into medicine. Women physicians have traditionally sought a more harmonious balance between work and family life.

Generational changes reflected societal changes and also contributed to evolving attitudes about work. The medical profession in T.E.B.'s time had consisted of members of the *Silent Generation*, born before 1946, and *Baby Boomers*, born between 1946 and 1964. Physicians from the *Silent*

Generation and *Baby Boomers* defined professionalism predominately in terms of hours worked and complete dedication to their job.

Members of *Generation X*, born between 1965 and 1980, and *Millennials*, born since 1980, are now the most populous group of physicians. They have grown up savvy and enthusiastic about technology but skeptical of entrenched cultural norms. They don't share their elders' preoccupation with work. They are less career-driven and demand flexible work hours.[321] Ultimately they evaluate success by the quality of their total life, not just by their work.[322]

This change in attitudes has produced a preference among many young physicians toward specialties that allow greater time for personal and family activities. For example, radiology, ophthalmology, anesthesiology, and dermatology, among others—denote high-paying fields with comfortable lifestyles.[323]

By 2000, the importance of lifestyle considerations for the younger generation unsettled many older physicians. The latter viewed younger doctors' desire to work fewer hours as *unprofessional*. Senior doctors had often belittled the qualities of younger generations. David Rogers pointed out: "Whenever physicians of middle age and beyond gather together ... one topic inevitably receives much hand-wringing attention—the sorry state of today's house staff."[324]

The rise of *lifestyle considerations*, however, has not really signified a decline of *professionalism* in medicine; it represents rather a new chapter in an ongoing story of generational change. *Generation X* and *Millennial* doctors are smart and energetic. They demonstrate a strong work ethic and a firm commitment to their medical careers. They define *professionalism* as excellence, not endurance. Some new forms of idealism have appeared, such as an interest in global health. No one wants doctors who don't believe they have a moral obligation to their patients. But no one wants exhausted or burned out doctors either. As the average work week in America has shortened, are not doctors, like anyone else, entitled to respite?[325]

Ludmerer summarized the predicament in his 2015 book, a statement which Brittingham would have supported:

> Developments in medical science bring us to the brink of unparalleled opportunity in preventing and treating

disease and relieving suffering ... The opportunity is there to envision medical education and practice as they should be, not as they are, and to work toward achieving that end. Such opportunities are to be treasured, not feared. The country will always need good doctors, and the medical profession has little to fear in the changes ahead as long as it remembers that it exists to serve, that the needs of patients come before its own.[326]

CHANGES IN MEDICAL PRACTICE

As endoscopy and imaging techniques such as CT and MRI scans simplified diagnosis in the late twentieth century, bedside examination skills became less important. Physicians and residents in internal medicine performed increasingly cursory physical examinations. They performed fewer routine pelvic and rectal exams. And they no longer habitually examined ear drums with otoscopes and retinae with ophthalmoscopes.

To be sure, there are many instances in which delivery of appropriate care can be rapid. But not all patients are served by speed and efficiency. The ones who suffer most are those with complicated diagnostic or treatment problems and those with significant psychological or social issues.

T.E.B. warned about the risk of relying totally on lab tests and X-rays. He was realistic about his wariness of changes, but adequate proof would overcome his doubts. Most of the doctors interviewed for this book thought T.E.B. would have accepted changes that were beneficial for patients, and he would have enjoyed looking in the endoscope with the endoscopist as much as he enjoyed doing a good physical exam. "T.E.B. would have loved CT and MRI scans," said Dr. Mayes. T.E.B. could make adjustments. He was not allergic to technology.

Today one might ask, "What's that stethoscope around your neck? Do you plan to be on TV?" T.E.B. would chuckle, "Well, you're a lot smarter than I, but I still get a lot of information from my stethoscope."[327]

According to Dr. Bo Sheller, a former Vanderbilt medical student and now professor of medicine in the division of pulmonary and critical care medicine, a portable ultrasound on a smartphone app will soon be in every physician's pocket. Every doctor will be able to do a basic bedside echocardiogram, scan the thyroid for nodules, determine the size of the liver and spleen, and examine the fetus in utero. Commenting on the electronic medical record, Sheller said that it has helped meet requirements for billing, but there is disagreement about its overall value: "T.E.B. taught students to look at the patient during a conversation. Looking at a computer screen while listening to a patient removes the effect of body language in the encounter." Many doctors are concerned that the extensive use of boilerplate templates is redundant and not informative. They long for the days when physicians wrote perceptive patient notes and did not just copy and paste a previous one with minor editing.[328]

In a recent article in the *New England Journal of Medicine*, Rosenthal and Verghese asked "What is the actual work of a physician?" They answered that the majority of work takes place away from the patient:

> Our attention is so frequently diverted from the lives, bodies, and souls of the people entrusted to our care ... We've distanced ourselves from the personhood of patients to do our work on the computer ... Meanwhile, drop-down menus, cut-and-paste text fields, and lists populated with a keystroke have created a medical record that at best reads like meaningless repetition of facts and at worst amounts to misleading inaccuracies.[329]

Changes in medical care have also resulted from expansion of large scale studies. In T.E.B.'s era there were many reasonable differences in opinion about diagnoses or treatments, and few good investigations with long follow-up were available to resolve the differences. In the 1970's, huge, expensive randomized clinical trials to resolve clinical questions were initiated. They led to *evidence-based-medicine,* and now many clinical decisions can be based on scientific evidence. Periodic analytic summaries and guidelines are available to inform good clinical care. Nevertheless, individual physicians are still vitally important to navigate the wealth of information and apply it to individual patients.[330]

Dr. Seth Cooper, chief resident in medicine in 1974-75 and now a retired hematologist, said that T.E.B. prepared his students for changes.

Humility prompted T.E.B.'s statement, "We don't know everything, and everything isn't known."[331]

THE PHYSICAL EXAM

Dr. Frank Boehm is a Vanderbilt medical graduate and currently vice-chairman of Obstetrics and Gynecology. He remembers an incident that happened fifty years ago, when he was a medical student, as though it were yesterday. He says he will remember it for the rest of his life. At 3 a.m, he worked up a patient who had entered the hospital coughing up blood. The next morning T.E.B. made rounds with a group of students. After Boehm presented his history and physical exam, Dr. Brittingham sat down at the bedside, his eyes level with those of the patient, and asked several questions. When the patient spoke, T.E.B. never interrupted, even though some of the patient's answers were lengthy and convoluted. Then Brittingham asked the patient to sit up, and he sat down on the bed behind the patient. Starting at the upper part of the right side of the patient's back, he tapped the back with the tip of his bent right third finger. He moved downward about an inch between each percussion. Every tap, produced by a sharp, forceful flexion of his right wrist, evoked a resonant, high-pitched sound—indicating normal air-filled lung—until he reached the middle of the chest. Suddenly, the percussion note changed to a dull sound. With a jubilant gleam in his eyes, T.E.B. turned to the students behind him and asked, "Did you hear that? Move closer so you can hear each percussion."

The students edged forward. Holding a ballpoint pen in his mouth, T.E.B. percussed from above down to the upper margin of the dull note, then drew an arc of blue ink on the patient's skin where the percussion sound changed to dullness. He repeated this maneuver starting at the spine, moving from left to right, and sketched a blue arc where the percussion note changed; then, starting at the right margin of the chest, moving from right to left; then at the bottom, advancing his percussion from below upward. The four arcs outlined an area of dullness in the middle of the right lung—indicating abnormal, non-aerated lung—about the diameter of a baseball. When the patient had a chest X-ray later that morning, a mass was evident in the right lung exactly where Dr. Brittingham had drawn his circle on the skin.[332]

Today one might consider that a chest X-ray done first would have made T.E.B.'s physical exam superfluous. A CT-scan of the chest, a bronchoscopy, and some laboratory tests would have supplied the information needed to diagnose and treat the patient with a high degree of precision. But, in T.E.B.'s era, doctors needed to listen to a patient's concerns, touch the patient, and establish a patient-physician bond. What is the significance of the patient-physician bond anyway? In my opinion, it plays a role in limiting medical malpractice lawsuits. But it is not fully known if it influences mortality, morbidity, or costs.

T.E.B.'s historical legacy was his skill in inspiring students to search for and use the best information available at the time. Every physician I questioned for this book shared this view. Acknowledging that we have progressively gained—and will continue to gain—technological expertise, T.E.B.'s historical legacy is analogous to that of Osler, who used the autopsy as a primary tool to learn and teach about disease a hundred years ago—and also to that of a more recent mentor like Dr. Grant Liddle, who used cumbersome rat bio-assays and laborious 24-hour collections of urine to learn and teach about the pituitary and adrenal glands.[333] It's impossible to compare doctors by the standards of different times. Civil War surgeons did not wash their hands. They worked without knowledge of the nature of infection and lacked drugs to treat it. They saved lives, using the best information available to them.[334]

INSURANCE AND FINANCES IN HEALTH CARE

After World War II, an admiring public bestowed abundant resources on the medical profession. By 1970, many Americans considered health care as a right. The majority had some type of medical insurance, and the number of charity patients had declined. But escalating costs changed the situation. Within two decades after enactment of Medicare and Medicaid in 1965, the cost of health care had risen from 4 percent of the gross domestic product to 11 percent. Today, health care accounts for more than 17% of gross domestic product. Rising costs did not always lead to increased benefits. The federal government, insurance companies, and equity investors insidiously gained control of health care.

To help control hospital costs, the government established the *DRG (diagnosis-related group)* system. The DRG program was well-intentioned but changed hospital culture. Hospitals received a fixed payment from insurers per case, depending on which DRG applied to the patient's condition. Therefore, to achieve financial success, hospitals sought shorter lengths of stay and a swifter turnover of patients. Sometimes it seemed that doctors were *processing* patients rather than thoughtfully treating them. Faculty at some teaching hospitals noted a decline in *reflective doctoring* and a less questioning attitude among residents.

Powerful political, economic, and cultural forces transformed medical practice into a commercial system—in ambulatory clinical sites as well as hospitals. With specialization, care became fragmented among several physicians, often with no one assuming responsibility for the patient. Direct-to-consumer TV advertising was a windfall for pharmaceutical companies, but the prices of many drugs skyrocketed. Some patients had batteries of tests before even seeing a physician. For example, a patient with back pain might have a costly MRI of the spine before seeing a neurosurgeon. Some now think that the business plan has become the driving force, not the welfare of patients.[335]

T.E.B. would not have welcomed or been able to stop this chain of events. But he would have continued treating his patients and teaching his students as this book has described. Sir William Osler addressed the issue a century earlier:

> As the practice of medicine is not a business, and can never be one, the education of the heart—the moral side of man—must keep pace with the education of the head. Our fellow creatures cannot be dealt with as man deals in corn and coal; 'the human heart by which we live' must control our professional relations.[336]

T.E.B.

THE MARVELS AND LIMITATIONS OF MODERN MEDICINE

Do the marvels of modern medicine—CT and MRI scans, ultrasound, robotic surgery—have an alternative point of view—the limitations of technology? Or have the fundamentals of medicine persisted?

The following story illustrates the limitations of technology and of shared physician responsibility. A patient in the intensive care unit at the V.A. Hospital was critically ill with bacterial endocarditis, an infection of the blood stream. Bacterial endocarditis results when a colony of bacteria infects and persists on a heart valve. In this patient's case it was the malfunctioning mitral valve. Numerous cultures of the blood had identified the responsible microorganism. Despite appropriate treatment with antibiotics, the patient had persistent shaking chills and fevers over two weeks. The doctors had considered the possibility of an abscess at the valve ring, where the valve leaflet attaches to the heart muscle, but an echocardiogram (ultrasound examination of the heart) had indicated no such lesion. On CT scans the only suspicious finding was an ill-defined defect in the spleen. The team of physicians included a cardiologist, a pulmonologist, a general surgeon, a thoracic surgeon, and all their residents and fellows. They were getting nowhere.

They called Dr. John Leonard, a specialist in infectious disease, for an opinion. Dr. Leonard reviewed the history and examined the patient. He conceded that he also did not know why the fever persisted. He recognized, however, that none of the specialists, each attending to his own area of expertise, had assumed primary responsibility for the patient's care. He decided to take charge, and he proceeded in a methodical fashion.

First, he called the general surgeon and requested him to remove the suspicious spleen. The surgeon responded that he could perform this surgery but really thought the operation had a low chance of resolving the illness. Leonard knew that when a surgeon expresses reluctance, it's unwise to push him.

Next, Dr. Leonard explored the medical literature. There he found that the majority of patients who had persistent fever after appropriate treatment for endocarditis did have an abscess at the valve ring. Sometimes it was difficult to find by a standard echocardiogram, but it was often visible on a transesophageal echocardiogram, an echocardiogram done with the ultrasound probe positioned in the esophagus, directly behind the heart. When this test was performed, it also showed no evidence of an abscess at the valve ring.

Leonard went to the literature again, this time to the radiology literature. Here he learned that transesophageal echocardiography had only 80% sensitivity in detecting a valve ring abscess. In other words, a transesophageal echocardiogram could be normal 20% of the time when such a lesion was actually present. Armed with this information, Leonard felt so confident of the diagnosis that he persuaded—yes, persuaded—the thoracic surgeon to operate. The surgeon found a valve ring abscess, drained it, and the patient got well.[337]

Was Leonard's tenacity just a coincidence? He had been a disciple of T.E.B. both as a medical student and as chief medical resident. After T.E.B. left Vanderbilt, Dr. Leonard assumed T.E.B.'s job as director of the medical residency program for the next twenty-one years. T.E.B. had left a lifelong impression on John Leonard.

Was the influence of Dr. Tom Brittingham just of historical significance for a generation of students and residents, a bright fire which has burned out and is no longer relevant?

If forced to choose, would informed, technology-infatuated Americans opt for a strong patient-physician bond, or prompt unlimited access to high-technology care, usually performed by a stranger?

Most patients don't want the choice. They want both. The challenge to the medical profession lies in the application of T.E.B.'s principles of doctoring in today's era of science and technology.

While a wave of new technology and profit-seeking rolled over medicine in the late twentieth century, it did not refute T.E.B.'s brand of doctoring. Brittingham was more than a voice in the wilderness; he was an exemplar of basic values in medicine. His example provided motivation and guidance to young doctors, who then strove to emulate him. Many felt he was still watching, and they did not want to disappoint him. He was a transitional figure, standing between the Oslerian tradition of doctoring and the more depersonalized era of modern medicine. Dr. Clif Cleaveland said, "His special gift to us was not to be found in facts about illness, but rather in making us aware of a process by which a sick person might be approached."[338]

Like his predecessor Dr. William Osler 75 years beforehand, T.E.B. was a teacher of internal medicine and a working doctor. Though some might say he considered each patient a research project, he was not an experimental scientist. He knew that medicine must rest on science and that medical scientists and their discoveries could amplify his effectiveness—but they could never replace the doctor seeing patients. His impact as an educator was to teach new physicians how to doctor by working at the bedside.

What does T.E.B.'s career tell us about the history of medicine? Does T.E.B.'s approach to doctoring challenge in any way the value of technological advances? Paradoxically, it may reveal how deficiencies he noted in his era could become a harbinger of improvements in the future. For example, he wrestled with the previous shortcomings in the diagnosis of functional disease, the treatment of cancer, and the treatment of coronary heart disease. In a positive sense, his career shows the importance of assuming total responsibility for the well-being of a patient—that of a doctor's comprehensive care versus care by multiple specialists with no one in charge. He cast light on the complementary—not divergent—roles of taking a history *and* a CT scan.

Dr. Alexander McLeod, a medical resident under Dr. Brittingham's tutelage and then a practitioner of general internal medicine, told me, "He was a man on fire, but not for the usual reasons. He had no career agenda, he was not climbing the professional pyramid. He was a constellation of stars that only happens once."[339]

By any measure, T.E.B. was a complex individual. One can wonder how his life was affected by the early death of his father, by the wealth he inherited, by his military experience in World War II. Was David Rogers a friend, an idol, a fairy godfather, or all three? We know that T.E.B. was a gifted and compassionate doctor, a loving son, husband, and father, with a ready smile and chuckle, an outgoing personality, a sharp sense of humor, and a few quirks.

Dean Chapman, a man with considerable perspective on faculty coming and going at Vanderbilt, asked Brittingham to reconsider his decision to leave academic medicine. Chapman observed that "a certain weariness sets in when you have completed what you set out to do." When Chapman asked Brittingham why he was leaving, T.E.B. said simply, "It's time."[340]

T.E.B.

Chief Residents in Medicine, Vanderbilt Hospital, 1963-1980

Lawrence K. Wolfe, M.D., 1963-64

Harold H. Sandstead, M.D., 1964-65

Richard L. Doyle, M.D., 1965-66

William H. Hall, Jr., M.D., 1966-67

Clifton R. Cleaveland, M.D., 1967-68

William Schaffner, M.D., 1968-69

Murray W. Smith, M.D., 1969-70

Charles E. Mayes, M.D., 1970-71

John S. Sergent, M.D., 1971-72

John M. Leonard, M.D., 1972-73

Burton C. West, M.D., 1973-74

R. Seth Cooper, M.D., 1974-75

R. Kirby Primm, M.D., 1975-76

Warren A. Hiatt, M.D., 1976-77

Marcus C. Houston, M.D., 1977-78

Michael B. Brenner, M.D., 1978-79

James E. Loyd, M.D., 1979-80

Bibliography

1. Michael Bliss. *William Osler—A Life in Medicine.* Oxford University Press. New York. 1999

2. Sallie Reynolds Matthews. *Interwoven: A Pioneer Chronicle.* University of Texas Press. Austin & London. 1936, republished 1958.

3. Frances Mayhugh Holden. *Lambshead before Interwoven: A Texas Range Chronicle, 1848 to 1878.* Texas A & M University Press. College Station. 1982.

4. S. C. Gwynne. *Empire of the Summer Moon.* Scribner. New York. 2010.

5. Laura Wilson. *Watt Matthews of Lambshead.* The Texas State Historical Association. Austin. 1989.

6. Kenneth M. Ludmerer. *Let Me Heal—The Opportunity to Preserve Excellence in American Medicine.* Oxford University Press. New York. 2015.

7. Lawrence K. Altman. *Who Goes First? The Story of Self-Experimentation in Medicine.* University of California Press. Berkeley, Los Angeles, London. 1986, 1987, 1988.

8. T. J. Stiles. *The First Tycoon—The Epic Life of Cornelius Vanderbilt.* Alfred A. Knopf. NewYork. 2009.

9. Rudolph H. Kampmeier. *Recollections—The Department of Medicine, Vanderbilt University School of Medicine, 1925-1959.* Vanderbilt University Press. Nashville. 1980.

10. Clif Cleaveland. *Sacred Space. Stories from a Life in Medicine.* American College of Physicians. Philadelphia. 1998.

11. William S. Stoney. *Pioneers of Cardiac Surgery.* Vanderbilt University Press. Nashville. 2008.

12. T.A. Preston. *Coronary Artery Surgery. A Critical Review.* Raven Press. New York. 1977.

13. Elisabeth Rosenthal. *An American Sickness—How healthcare became big business and how you can take it back.* Penguin Books, an imprint of Penguin Random House. New York. 2017.

14. David Oshinsky. *Bellevue—Three Centuries of Medicine and Mayhem at America's Most Storied Hospital.* Doubleday, a division of Penguin Random House LLC. New York. 2016.

Acknowledgments

Writing this book was an adventure, but I could not have accomplished it without the help of family, friends, and colleagues. It began one evening in 2015 as my wife Edie and I were eating dinner with my cousins, Stan and Sara Graber. Sara, a friend of Dotsy Brittingham for several years, nonchalantly mentioned that Dotsy wished someone would write something about her husband Tom, who had died twenty-nine years previously. After reflection, I decided to take on this challenging project. Sara and Stan arranged a lunch date for me to meet Dotsy, and the result was this book.

I was unaware then that Dr. Eric Dyer, a former Vanderbilt medical student and house officer who was intrigued with Dr. Brittingham, had spent five years on a similar project. While engaged in a full-time pulmonology practice, Eric had undertaken extensive research on T.E.B. He had interviewed dozens of former students and house officers, had traveled to all the sites where Brittingham had lived and worked, and had drafted a manuscript of T.E.B.'s biography, which he entitled *Brittingham: A Life in Medicine*. Dissatisfied with the preliminary version, he had started a revision, before his sudden and unexpected death in 2004. His widow, Cheryl Dyer, graciously made Eric's manuscripts and photographs available to me, in addition to Dr. Brittingham's extensive letters, which dated from age fourteen. Much of the research cited herein represents Eric's work. His contributions have significantly strengthened this book.

Patsy Des Prez provided detailed information about life in the Vanderbilt Department of Medicine when Brittingham joined the faculty. She lent me books (listed in bibliography) about the Texas range in the mid-nineteenth century and the Reynolds and Matthews families who established the Lambshead Ranch there. Thereafter, she referred me to Betty Collins, who in turn referred me to James J. Thweatt, Health

Information Specialist at the Vanderbilt Eskind Biomedical Library. Thweatt furnished invaluable assistance in locating Cheryl Dyer. He and his associate Chris Ryland in Special Collections offered their expertise and consultation dealing with the library's collections of Dr. Brittingham and of Dr. David Rogers. Laura Frater, head librarian at the Harold H. Brittingham Memorial Library at MetroHealth Medical Center in Cleveland, Ohio, supplied me with information about the dedication of that library in memory of T.E.B.'s father.

Dr. Thomas E. Brittingham IV provided the photo of his father in army uniform, and Drs. Clark Gregg and Susan Brittingham provided the wedding photo of T.E.B. and Dotsy. John Brittingham, T.E.B.'s oldest son, assisted me in acquiring copyright to reproduce Dr. Brittingham's writings, lectures, and letters. Laura Wilson, author and photographer of *Watt Matthews of Lambshead*, kindly granted me permission to reproduce two of the photos from her magnificent book.

During creation of this book, many doctors and others have shared anecdotes about T.E.B. or have participated in taped interviews with me. They include, in alphabetical order, Dr. Jean Ballinger, Dr. Bart Campbell, Dr. David Barton, Dr. Frank Boehm, Dr. Thomas E. Brittingham IV, Margaret Brittingham, Dr. Charles Bryan, Dr. Dan Canale, Dr. Fred Callahan, Dr. Clif Cleaveland, Dr. Alan Cohen, Dr. Seth Cooper, Dr. Rick Davidson, Dr. Tom Davis, Cindy Diamond, RN, Dr. John Dixon, Dr. David Dodson, Dr. Robert Dunkerly, Dr. Agnes Fogo, Dr. Frank Gluck, Dr. Stan Graber, Dr. Clark Gregg, Dr. David Gregory, Dr. Mark Houston, Dr. Brevard Haynes, Dr. Hank Jennings, Dr. Allen Kaiser, Dr. Sandy Krantz, Dr. Liz Kruger, Dr. John Leonard, Dr. James Loyd, Dr. Phil Majerus, Dr. Charles Mayes, Dr. Alex McLeod, Dr. Clifton Meador, Dr. Bill Mitchell, Dr. Ed Nelson, Dr. David Orth, Dr. Harry Page, Dr. Bill Petrie, Dr. David Robertson, Dr. Howard Rosen, Dr. John Sergent, Dr. William Schaffner, Dr. Bo Sheller, Dr. Jim Snell, Dr. Bill Stone, Dr. Bill Stoney, Dr. Jimmy Sullivan, Dr. Carl Vandevender, Dr. Carl Wierum, Dr. Larry Wolfe, and Dr. Taylor Wray.

Dr. Jimmy Sullivan gave me his copy of Kenneth Ludmerer's book, *Let Me Heal*, which methodically examined the development of the American residency system. Dr. Carl Mitchell provided me a copy of the extensive journal he kept as a medical student at Washington University and as an intern at St. Louis City Hospital, contributing significantly to Chapter 9. Dr. James Snell introduced me to Dr. Rudolph Kampmeier's *Recollections—The Department of Medicine, Vanderbilt University School of Medicine, 1925-1959.* Dr. Taylor Wray supplied me with a typescript of Dr. Brittingham's lecture on functional disease, reproduced in Chapter 13. Dr. John Sergent provided a copy of his acceptance speech on his receipt of the Distinguished Medical Educator Award at the Association of Resident Directors in Internal Medicine, which dealt with problem-solving in the Vanderbilt residency program. Dr. Alan Cohen provided me a copy of the letter which Dr. Brittingham wrote to the Koenig family and to the residents who cared for Dr. Glen Koenig during his illness. Many other former students and residents shared copies of letters they had received from Dr. Brittingham. Dr. Charles Bryan shared his knowledge of the legacy of Dr. William Osler, his perspective on the past and future of internal medicine, and a letter he received from Dr. Brittingham at the onset of T.E.B.'s terminal illness, requesting a copy of Bryan's article about cytomegalovirus infection as a cause of fever of unknown origin.

I want to thank the members of my writing club, the Scribblers—Barry Jones, John Davis, Robin Andrews, Anne Lane, Jillyn McCullough, Jerry Henderson, Clifton Meador, and Frank Freemon—for their support and their penetrating critiques of the chapters we discussed. Drs. Meador and Freemon subsequently evaluated the entire manuscript.

My editor and former creative writing teacher, Katie Hanson, reviewed the manuscript meticulously and offered indispensable suggestions, gently but firmly, for improving the narrative. After incorporating Katie's advice, I asked my friend Dr. Stephen Dummer, a former journalist and retired Vanderbilt physician, for one last critical appraisal. Steve read the manuscript three times and offered many vital suggestions.

At home, my son Dean and my wife Edie evaluated the manuscript. Their critical reviews challenged and grounded my writing. Edie tolerated my immersion in this project and my attention to the computer screen with the patience and understanding she has always shown throughout our fifty-nine year marriage.

Notes

1. Bliss, Michael. *William Osler: A Life in Medicine.* Oxford University Press. New York. *2007*, William Osler, one of the fathers of modern medicine, was a brilliant, innovative teacher who revolutionized the art of practicing medicine at the bedside of his patients. He was idolized by two generations of medical students and practitioners for whom he came to personify the ideal doctor. But much more than a physician, Osler was a fiercely intelligent humanist.
2. Dr. Henry (Hank) Jennings, interview by author, Nashville, Tennessee, April 29, 2017. Description of chairs in T.E.B.'s office. Dr. Jennings was a student of T.E.B. in medical school and on the house staff and is currently an interventional cardiologist at Vanderbilt Hospital.
3. Dr. Grant W. Liddle, Chairman of the Vanderbilt Department of Medicine, 1963-1982.
4. Dr. James Haynes, interviews by Dr. Eric Dyer, Brentwood, Tennessee, November 14, 2002 and February 2, 2003. Haynes successfully completed internship, residency and fellowship at Vanderbilt, and enjoyed a distinguished career practicing pulmonary and critical medicine at St. Thomas Hospital in Nashville.
5. Dr. Judson Rogers, interview by Dr. Eric Dyer, Nashville, Tennessee, November 13.2002. Dr. Rogers practices internal medicine in Nashville.
6. Dr. John (Dick) Dixon, interview by author, Nashville, Tennessee, April 1, 2016. Dr. Dixon is a cardiologist at Vanderbilt Medical Center.
7. Bliss, Michael. *William Osler,* 3
8. Sallie Reynolds Matthews. Interwoven: A Pioneer Chronicle. University of Texas Press. 1936, republished 1958. Austin & London, 118-121
9. Ibid., 119
10. Ibid., 4-6
11. Ibid., 23
12. Ibid., 8
13. Ibid., 33-34
14. Ibid., 34-37
15. Ibid., 111

16. Frances Mayhugh Holden. Lambshead before Interwoven: A Texas Range Chronicle, 1848 to 1878. Texas A & M University Press: College Station. 1982, 111-157
17. Sallie Reynolds Matthews. Interwoven, 141
18. Texas State Historical Association. Mackenzie, Ranald Slidell. https://tshaonline.org/handbook/online/articles/fma07. (Accessed January 20, 2016)
19. Gwynne, SC. Empire of the Summer Moon. Scribner. New York, 2010, 255
20. Sallie Reynolds Matthews. Interwoven, 184
21. Frances Mayhugh Holden. Lambshead before Interwoven, 9-18
22. Ibid., 20-28
23. Eric Dyer. "Brittingham: A Life in Medicine." (Unpublished manuscript), 2005.
24. Sallie Reynolds Matthews. Interwoven, Introduction, x
25. Eric Dyer. (Unpublished manuscript), 2005.
26. Jill J. Dunkel. Watt Matthews: Dean of Texas Cattlemen. American Cowboy. 2014
27. Ibid.
28. Laura Wilson. Watt Matthews of Lambshead. The Texas State Historical Association, Austin. 1989, 62
29. Ibid., 42.
30. Robert McG Thomas. Watkins Matthews, Rancher from Bygone Era, Dies at 98. Obituary, New York Times, 1997
31. ancestry.com
32. St. Louis Globe Democrat, July 1, 1900
33. Eric L. Dyer. "Brittingham—A life in Medicine." (Unpublished manuscript), 2005
34. Mark Twain. *Mississippi Writings. Chronology.* New York. 1982, 1057-1061.
35. Paine, Albert Bigelow. *Mark Twain, a Biography*. Volume 1. Harper and Brothers, New York. 1912, 77
36. Ibid., 90
37. UW archives and records management. www.library.wisc.edu/archives/exhibits/campus-history-projects/the-brittingham/family/thomas-evans/brittingham/. Accessed January 7, 2016
38. http://madisonparksfoundation.org/thomas-e-brittingham, accessed September 6, 2017
39. Ibid.
40. Thomas Evans Brittingham (grandfather of the subject of this book) to Mr. Juan F. Brittingham, Gomez Palacio Durango, July 25, 1906
41. David K. Ford. (Unpublished manuscript), July, 1959
42. Ludmerer, Kenneth K. *Let me Heal—The Opportunity to Preserve Excellence in American Medicine*. Oxford University Press. New York. 2015, 111
43. Ibid., p. 103.

44. Dorothy Mott Brittingham, interview by Dr. Eric Dyer, Nashville, Tennessee, July 14, 2004
45. Ibid, January 8, 2003
46. Watt Matthews, letter to Lucille Matthews, 23 Sept, 1921
47. Thoracoplasty is a surgical procedure which collapses the diseased lung and the cavities of pulmonary tuberculosis by removing or resecting one or more ribs. The operation was designed to give the lung a chance to rest and repair itself and to cut off the oxygen supply and therefore kill the tuberculosis bacteria.
48. Down Syndrome is a chromosomal abnormality, not inherited, which occurs as a random event during cell division in fetal development. It results in mild to moderate intellectual disability, a characteristic facial appearance with an upward slant of the eyes, and short stature. One in every 691 babies in the United States is born with Down syndrome, making Down syndrome the most common genetic condition. Approximately 400,000 Americans have Down syndrome and about 6,000 babies with Down syndrome are born in the United States each year. National Down Syndrome Society. https://www.ndss.org, accessed November 21, 2016
49. Commissurotomy is an open-heart surgical repair of a mitral valve that is narrowed by mitral stenosis, usually a consequence of rheumatic heart disease.
50. Robert. N. Buchanan, Jr., interview with Dr. Eric Dyer, Nashville, TN, December 3, 2002
51. Hawken School, March, 1936, Character Report of Tom Brittingham
52. Thomas E. Brittingham II, letter to Harold H. Brittingham, Shaker Heights, Ohio, July 7, 1936
53. Ethyl Matthews Casey, Lucile's older sister, from San Antonio, Texas
54. Margaret Brittingham Reid, Harold's sister, from Lake Forest, Illinois
55. Watt Matthews, Lucile's brother, from Albany, Texas
56. Thomas Evans Brittingham, Jr., Harold's brother
57. Margaret Brittingham, Thomas Evans Brittingham, Jr.'s wife.
58. Stecher, Robert M. *Story of a Library.The Harold H. Brittingham Memorial Library at Metropolitan General Hospital,Cleveland, Ohio.* Bulletin of the Cleveland Medical Library, Vol XV, No.a, January, 1968, 4-13. Provided by Laura Frater, head librarian, August 4, 2016
59. Laura Frater, head librarian at The Harold H. Brittingham Memorial Library. The library is located on the second floor (Room 267) of the Rammelkamp Center for Education and Research at MetroHealth Medical Center. 2500 Metrohealth Drive, Cleveland 44109 The library hours are Monday through Friday, 9 a.m. - 10 p.m.; Saturday, 9 a.m. - 5 p.m., and Sunday, 1 - 5:30 p.m. The library phone number is 216-778-5623 or -4313
60. David K. Ford,(Unpublished manuscript) July, 1959
61. Donald M. Bell, MD, to Thomas E. Brittingham II, Cleveland, Ohio, March 22, 1937

62. Hawken School, March, 1938, Character Report of Tom Brittingham
63. Betty Densmore, interview with Dr. Eric Dyer, Albany, Texas, June, 2003
64. Feller was a pitcher for the Indians from 1936 to 1956, pitched three no-hitters during his career, and was inducted into the Hall of Fame. In October, 1938, he established a Major League record by striking out eighteen Detroit Tigers in a nine-inning game
65. Sally Ann Judd Harrison, interview by Dr. Eric Dyer, Wharton, Texas, August 19, 2004
66. Kolowrat, Ernest. *What Made Maria Do it?*, in *A Chronicle of an American School*. New Amsterdam Books. New York, 1992
67. Ibid
68. T.E.B. from Hotchkiss. School, Lakeville, Connecticut to John M. Brittingham, San Antonio, Texas, December 10, 1938
69. T.E.B. from Hotchkiss School, Lakeville, Connecticut, to Robert Brittingham, San Antonio, Texas, October 30, 1939
70. Don Durgin and Tom Brittingham continued as roommates at Princeton. Durgin eventually became President of NBC Television and a Director of Dunn and Bradstreet
71. T.E.B. from Hotchkiss School, Lakeville, Connecticut to Ms. H. H. Brittingham, San Antonio, Texas, January 8, 1939
72. Ibid., May 7, 1939
73. Ibid., February 18, 1940
74. Ibid, January 28, 1941
75. Ibid., April 27, 1941
76. Ibid., May 20, 1941
77. T.E.B., St. Louis, Missouri, to Mrs. H.H. Brittingham, Fort Worth, Texas, February 28, 1959
78. T.E.B., Camp Roberts, California, telegram to Mrs. H.H. Brittingham, Fort Worth, Texas, March 24, 1943
79. Dr. Eric Dyer. (Unpublished manuscript). 2005
80. Dr. Charles Mayes. Interview by author, Nashville, Tennessee, May 12, 2016
81. Dr. Thomas E. Brittingham IV. Interview by author, Nashville, Tennessee, August 13, 2017. Thomas E. Brittingham IV, T.E.B.'s youngest son, worked as an electrical engineer for 16 years, before entering Vanderbilt Medical School in 1995
82. T.E.B., Nashville, Tennessee, to Susan Brittingham, March 12, 1973
83. Eric Dyer. (Unpublished manuscript). 2005
84. Dorothy Mott Brittingham. Interview by author, Nashville, Tennessee, February 20, 2018
85. T.E.B., Boston, Massachusetts to Ms. H. H. Brittingham, Fort Worth, Texas, September 28, 1946
86. Ibid., September 21, 1946

87   Ibid.
88   Ibid., October 6, 1946.
89   T.E.B., Nashville, Tennessee, to Susan Brittingham, June 3, 1973
90   T.E.B., Boston, Massachusetts, to Ms. H. H. Brittingham, Fort Worth, Texas,, October 25, 1946
91   Ibid., November 3, 1946
92   Dorothy Mott Brittingham, interview with Dr. Eric Dyer, Nashville, Tennessee, January 22, 2003.
93   T.E.B., Boston, Massachusetts, to Ms. H. H. Brittingham, Fort Worth, Texas, October 26, 1946
94   Ibid., January 5, 1947
95   Ibid., February 23, 1947
96   Ibid., April 6, 1947
97   Ibid., September 28, 1947
98   John Burnum, MD, interview with Dr. Eric Dyer, Tuscaloosa, Alabama, February 27, 2003
99   T.E.B., Boston, Massachusetts to Ms. H. H. Brittingham, Fort Worth, Texas, February 9, 1947
100   Ibid., April 20, 1947
101   Ibid., September 28, 1947
102   Dorothy Mott Brittingham, interview with Dr. Eric Dyer, Nashville, Tennessee, January 22, 2003
103   Eric L. Dyer. *Brittingham—A Life in Medicine*. (Unpublished manuscript), 2008.
104   T.E.B., Boston Massachusetts to Dorothy S. Mott, Strobl, Austria, August 2, 1948
105   Ibid., to Dorothy S. Mott, London, England, August 24, 1948
106   Dorothy Mott Brittingham, interview with Dr. Eric Dyer, Nashville, Tennessee, February 5,   2003
107   Dorothy Mott Brittingham, Interview with author, Nashville, Tennessee, February 6, 2017
108   Ibid.
109   Ibid.
110   Ludmerer, KM. *Let Me Heal—The Opportunity to Preserve Excellence in American Medicine.* Oxford University Press. New York. 2015, 17-36
111   Ibid., 73
112   Ibid., 85
113   Ibid., 92
114   T.E.B., New York, New York, to Mrs. D. M. Brittingham, Fort Worth, Texas, October 13, 1950
115   Dorothy Mott Brittingham, Interview with author, Nashville, Tennessee, February 20, 2017
116   Ibid, interview with Dr. Eric Dyer, Nashville, Tennessee, January 29, 2003

117 S.J. et al. *The 100 most eminent psychologists of the 20th century*. Review of General Psychology. 2002, 6: 139–152
118 Rogers, Carl. *On becoming a person: A therapist's view of psychotherapy*. Constable. London. 1961
119 Dr. Carl Wierum, Demarest, New Jersey, telephone interview with author, May 2, 2015
120 Dr. David Rogers to T.E.B., February 27, 1953
121 Des Prez, R. *A Tribute to Dr. Thomas E. Brittingham*. Vanderbilt Medicine. Winter, 1987
122 Eric Dyer. "Brittingham—A Life in Medicine." (Unpublished manuscript), 2005
123 Altman, Lawrence K. *Who Goes First? The Study of Self-Experimentation in Medicine*. University of California Press. Berkeley & Los Angeles. 1986, 282
124 T.E.B., St. Louis, Missouri, to Dorothy M. Brittingham, Fort Worth, Texas, November 1, 1954
125 Altman, 286
126 Ibid., 291
127 T.E.B., St. Louis, Missouri, to Dorothy Not Brittingham, Albany, Texas, March 16 to March 19, 1957
128 Dr. Carl Moore, St. Louis, Missouri, to Dr., Samuel D. Martin, Gainesville, Florida, February 27, 1958
129 Altman, 295
130 Dr. Carl Mitchell. Interview with author. Nashville, Tennessee, September 10, 2015
131 Ibid
132 T.E.B., St. Louis, Missouri, to Dorothy M. Brittingham, Brightwaters, New York, July 5, 1959
133 Dr. Carl Mitchell. Interview with author. Nashville, Tennessee, September 10, 2015
134 Ibid
135 T.E.B., St. Louis, Missouri, to Mrs. H.H. Brittingham, Fort Worth, Texas, October 24, 1961
136 Dr. Carl Mitchell. Interview with author. Nashville, Tennessee, September 10, 2015
137 T.E.B., St. Louis, Missouri, to Dorothy M. Brittingham, Quogue, Long Island, August 7, 1960
138 T.E.B., St. Louis, Missouri, to Ms. H. H. Brittingham, Fort Worth, Texas, May 10, 1962
139 T.E.B. to Dr. Carl V. Moore, June 20, 1961
140 Dr. Carl V. Moore, St. Louis, Missouri, to Dr. David E. Rogers, Nashville, Tennessee, June 19, 1962
141 T.E.B., St. Louis, Missouri, to Ms. H. H. Brittingham, Fort Worth, Texas,. January 20, 1962

[142] Stiles, T. J. *The First Tycoon—The Epic Life of Cornelius Vanderbilt*. Alfred A. Knopf. NewYork. 2009, 531.

[143] Kampmeier, R.H. *Recollections—The Department of Medicine, Vanderbilt University, 1925-1959*. Vanderbilt University Press. Nashville. 1980, 1-13.

[144] Ibid., Appendix 1. p. 283. In an open letter to President Remsen of Johns Hopkins University in 1911, Osler wrote: "The subject of whole-time clinical teachers ... is a big question, with two sides. I have tried to see both, as I have lived both ... and can appreciate both." He conceived of full-time clinical professors ... "usually of quiet studious habit, not built for battle ... who have been content to work solely at the problems of disease. How shall such a man, 'cabined, cribbed, confined' within the four walls of a hospital practicing the fugitive and cloistered virtues of a clinical monk, how shall he, forsooth, train men for a race, the dust and heat of which he knows nothing and—this is a possibility!—care less? ... The danger would be the evolution throughout the country of a set of clinical prigs, the boundary of whose horizons would be the laboratory, and whose only human interest was research, forgetful of the wider claims of the clinical professor as a trainer of the young, a leader in the multiform activities of the profession and interpreter of science to his generation, and a counselor in public and private to the people in whose interest after all the school exists."

[145] Dr. C. Sidney Burwell had been a cardiac research fellow with Dr. Paul Dudley White and had worked on methods of determining the cardiac output. He was on the faculty of Vanderbilt Medical School from 1925 to 1935, the last seven years as Head of the Department of Medicine. He left Vanderbilt to become dean of his alma mater, Harvard Medical School.

[146] Dr. Tinsley Harrison, a seventh generation physician, was considered the intellectual descendent of Sir William Osler, his father's friend and inspiration. He came to Vanderbilt in 1925 as Chief Resident in Medicine. In 1935 he published his first book, *Failure of the Circulation*, which he considered his best publication, although throughout his later life he protested vigorously that he was a general internist, not a cardiologist. Before age fifty he wrote what became a classic in medical literature, *Harrison's Principles of Internal Medicine*, translated into at least nine different languages. After leaving Vanderbilt in 1941 he started the Departments of Internal Medicine at Bowman Gray Medical School in Winston-Salem, North Carolina; at the Southwest Medical School in Dallas, Texas; and at the University of Alabama in Birmingham, Alabama. Later in his career he became a vocal critic of excessive technology and what he viewed as a turning away from the bedside history and physical examination. But when he had his second and fatal myocardial infarction (heart attack) in 1978, he was persuaded to go to the hospital only by the temptation of seeing the new thallium myocardial scan of his infarct. Then he insisted on returning home, where he died quietly in his own bed.

147. Dr. Rudolph Kampmeier joined the Vanderbilt medical faculty in 1936. He taught the course in physical diagnosis to the sophomore class. His textbook, *Physical Diagnosis in Health and Disease,* was the leading book on the subject at that time. He also wrote a major textbook on syphilis, *Essentials of Syphilology*. Syphilis was so prevalent that Kampmeier ran a special syphilis clinic at Vanderbilt. Dr. William Osler had said, "To know syphilis in all its manifestations is to know medicine."

148. Dr. Hugh Morgan, a tall, stately man who previously had an active private practice in Nashville and was on the Vanderbilt part-time faculty, became chairman of the department of medicine in 1935, a position he held until his retirement for failing health in 1958. While a college student at Vanderbilt, he had been an all-star center in football.

149. The most active and dependable part-time faculty members included Drs. Edgar Jones, Sam Riven, David Strayhorn, Clarence Thomas, and Albert Weinstein.

150. Dr. John Youmans was a full-time member of the Department of Medicine from 1927, two years after the new school opened, until he resigned to serve in the Armed Services in World War II. After the war he was dean of the University of Illinois College of Medicine until he returned to Vanderbilt School of Medicine as dean from 1950 to 1958.

151. Dr. John Chapman, Dean of Vanderbilt School of Medicine 1976-2001. Interview by Andrew White, 2nd year medical student, Nashville, TN, April 3, 2002.

152. Dr. Elliot Newman was an associate professor of medicine at Johns Hopkins when Dr. Morgan invited him to join the Vanderbilt faculty in 1952 as the first Joe and Morris Werthan Professor of Experimental Medicine and director of the Laboratory of Clinical Physiology, positions he held until his death in 1973. His effectiveness led to the establishment of the NIH-supported Clinical Research Center (CRC), which he built to national prominence within a few years.

153. Dr. Grant W. Liddle joined the Vanderbilt Department of Medicine in 1956 as chief of the Division of Endocrinology, after three years as an established investigator at the National Heart Institute. During his first three years at Vanderbilt he produced a dozen publications dealing with testing of pituitary-adrenal function. After the departure of David Rogers in 1968, Liddle became chairman of the Department of Medicine. His research in endocrinology resulted in international recognition of Vanderbilt as a center of endocrinology training and research.

154. David Rogers graduated from Cornell University Medical College, did his internship and residency training at Johns Hopkins, then returned to Cornell. There he was a fellow in infectious disease and then medical chief resident. He was recruited to Vanderbilt in 1959. Subsequently he was Dean of the School of Medicine at Johns Hopkins. From 1972 until 1987, he was the first president

of the $1.2 billion Robert Wood Johnson Foundation, the largest foundation in the United States devoted exclusively to health. His major projects focused on improving the delivery of health care, particularly to the poor and to racial minority groups. When he retired from the Foundation, he returned to Cornell to a chair endowed by his mentor, Dr. Walsh McDermott. At the end of his career, he became a major advisor for the New York State AIDS Advisory Council and Vice-Chair of the President's National Commission on AIDS. He was responsible for the national guidelines on AIDS policy.

[155] Waddle, R. *Days of Thunder—The Lawson Affair.* Vanderbilt Magazine, Fall 2002

[156] Dr. James Snell. Interview by author, Nashville, Tennessee, February 27, 2017.

[157] Dr. William Schaffner. Interview by author, Nashville, Tennessee, February 16, 2016. Schaffner was chief resident in medicine in 1968-69 and currently is Professor of Medicine and past Chairman of the Department of Preventive Medicine at Vanderbilt.

[158] Dr. Charles Mayes. Interview by author. Nashville, TN, May 12, 2016. Mayes is now a retired cardiologist in Nashville.

[159] Dr. Agnes Fogo. Interview by author. Nashville, Tennessee, December 30, 2015. Dr. Fogo is now a Professor of Pathology at Vanderbilt.

[160] Des Prez, R. *A Tribute to Dr. Thomas E. Brittingham.* Vanderbilt Medicine, Nashville, TN, Winter 1987.

[161] Dr. Mark Houston. Interview by author, Nashville, Tennessee, December 28, 2015. Houston was chief resident in medicine in 1977-78.

[162] Dr. Marvin Gregory. Interview by Dr. Eric Dyer, Nashville, Tennessee, January 23, 2003. Gregory graduated from Vanderbilt medical student in 1966.

[163] Roberts, Owen W. *Class of 1921 Notes.* Princeton Alumni Weekly, June 4, 1997

[164] This guide is a revision of a similar document T.E.B. prepared at St. Louis City Hospital four years earlier. This author received one as a third year Washington University medical student.

[165] This teaching session was discussed by Dr. Roger Des Prez at the dedication of the Brittingham Learning Center in 1993, seven years after T.E.B.'s death.

[166] Sergent, John S. The Nashville Tennessean, August 4, 1993.

[167] Dr. Rick Davidson. Telephone interview by author. Davidson graduated from Vanderbilt Medical School in 1972 and served his internship and one year of residency there. When he retired in 2013, he was Associate Vice President for Health Affairs at the University of Florida.

[168] Dr. William Stone. Interview by author, October 1, 2015. Nashville, Tennessee. Stone was on the Vanderbilt house staff in the early 1960s and has been chief of nephrology at Nashville Veterans Hospital for many years.

[169] Dr. Nace graduated from medical school, completed internal medicine residency and fellowships in nephrology and clinical pharmacology, all at Vanderbilt. He currently serves as Assistant Dean for Curriculum Integration at University of Tennessee Health Sciences Center in Memphis,TN

170. Dr. Cleaveland was a Rhodes Scholar before earning his medical degree from Johns Hopkins University. He was chief medical resident under Dr. Brittingham at Vanderbilt University Hospital in 1967-68. He practiced internal medicine 35 years in Chattanooga, Tennessee, until his retirement in 2004. He is a former president of the American College of Physicians.
171. Dr. James E. Loyd. Interview by author, Nashville, Tennessee, December 30, 2015. Dr. Loyd trained in Internal Medicine at Vanderbilt from 1973-1978 and was the last chief resident in medicine under Dr. Brittingham. He is now Professor of Medicine in the Division of Allergy, Pulmonology, and Critical Care Medicine at Vanderbilt.
172. Dr. John Sergent. Interview by author, Nashville, Tennessee, October 14, 2015. Dr. Sergent was Chief Medical Resident 1971-72 and Director of Medical Residency Program 2003-2013.
173. Dr. David E. Rogers to T.E.B., Nashville, TN, December, 1963.
174. T.E.B., Nashville, TN, to Lucile Matthews, Fort Worth, Texas, February 9, 1964
175. Ibid., October 4, 1964
176. Dr. Karl Vandevender, interview by author, Nashville, Tennessee, January 19, 2016. Dr. Vandevender was on the Vanderbilt house staff from 1979 to 1982 and currently practices internal medicine in Nashville.
177. T.E.B., Nashville, TN, to Susan Brittingham, Stanford University, Palo Alto, California, March 19, 1973
178. Eric Dyer. *Brittingham—A Life in Medicine." Unpublished manuscript), 2005
179. Dr. Frederick Callahan, interview by Dr. Eric Dyer, Nashville, Tennessee, November 24, 2002. Dr. Callahan currently practices neurology in Nashville, Tennessee
180. Dr. Dan Canale, interview by author, Nashville, Tennessee, May 31, 2016
181. T.E.B., Nashville, TN, to Lucile Matthews, Fort Worth, Texas, August 23, 1964
182. Dr. Eric Dyer. *Brittingham—A Life in Medicine." Unpublished manuscript), 2005
183. T.E.B., Nashville, Tennessee, to a failed intern, July 8, 1973.
184. Dr. Liz Kruger, interview by author, Nashville, Tennessee, July 13, 2017. Dr. Kruger graduated from Vanderbilt Medical School in 1979, trained in pediatrics and neonatology at Vanderbilt, then practiced thirty years as a neonatologist at Baptist Hospital in Nashville.
185. Dr. Robert Faber, interview by Dr. Eric Dyer, Nashville, Tennessee, February 28, 2003. Dr. Faber is a retired urologist in Nashville.
186. Dr. Charles Mayes, interview by author Nashville, Tennessee, May 12, 2016. Dr. Mayes is now a retired cardiologist in Nashville.
187. Dr. John Dixon, interview by author, Nashville, Tennessee, April 1, 2016.
188. Dr. Eric Dyer. *Brittingham—A Life in Medicine." Unpublished manuscript), 2005

189 Dr. Lawrence Wolfe, interview by author, Nashville, TN, September 9, 2015. Dr. Wolfe was chief resident in medicine at Vanderbilt in 1963-64. After that, he enjoyed a distinguished career practicing internal medicine and endocrinology in Nashville.

190 Dr. Robert Dunkerly, interview by author, Nashville, Tennessee, July 6, 2016. Dr. Dunkerly is a retired gastroenterologist in Nashville.

191 Dr. John Sergent, interview by author, Nashville, Tennessee, October 14, 2015.

192 Dr. James R. (Bo) Sheller, interview by author, Nashville, Tennessee, October 15, 2015. When Sheller graduated from Vanderbilt Medical School, T.E.B. wrote a letter to the University of California at San Francisco (UCSF), which Sheller says resulted In his acceptance there for postgraduate training. Eight years later he returned to Vanderbilt, where he is a professor of medicine in the division of pulmonary and critical care medicine.

193 Dr. Rick Davidson, Gainesville, Florida, telephone interview by author, 2016

194 Dr. David Gregory, interview by author, Nashville, Tennessee, February 8, 2016.

195 Dean John E. Chapman, interview by Dr. Eric Dyer, Nashville,Tennessee, March. 19, 2003. Dr. John E. Chapman was appointed Dean of Vanderbilt University School of Medicine in 1967 and served in that capacity with distinction for more than twenty-five years.

196 Bryan, C. *Caring carefully: Sir William Osler on the issue of competence vs compassion in medicine.* Baylor University Medical Proceedings 1999:12, 284

197 Dr. David Gregory, interview by author, Nashville, Tennessee, February 8, 2016

198 Ms. Elizabeth M. Hoppe, interview by Dr. Eric Dyer, Nashville, Tennessee, October 7, 2003

199 Dr. Mark Averbuch, interview by Dr. Eric Dyer, Nashville, Tennessee, November 23, 2002. Dr. Averbuch is now a retired internist in Nashville.

200 Dr. Dan Canale, interview by author, Nashville, Tennessee, May 31, 2016

201 Dr. John (Dick) Dixon, interview by author, Nashville, Tennessee, April 1, 2016.

202 Dean John E. Chapman, interview by Dr. Eric Dyer, Nashville, Tennessee, March 19, 2003

203 Dr. Liz Kruger, interview by author, Nashville, Tennessee, July 13, 2017.

204 Dr. Gary Duncan, interview by Dr. Eric Dyer, Nashville, Tennessee, February 4, 2003

205 T.E.B., Nashville, Tennessee, to Susan Brittingham, Stanford, California, November 7, 1971

206 T.E.B., Nashville, Tennessee, to Margaret Brittingham, Wellesley, Massachusetts, September 23, 1973

207 Dorothy Mott Brittingham, interview by Dr. Eric Dyer, Nashville, Tennessee, February 12, 2003

208 T.E.B., Nashville, Tennessee, to Susan Brittingham, Mexico, March 11, 1974

209 Ibid.

210 T.E.B. to Susan Brittingham, February 17, 1971
211 T.E.B. to Margaret Brittingham, November 20, 1974
212 Dr. Taylor Wray provided the author with a typescript of this lecture. Dr. Wray was a medical resident at Vanderbilt from 1969-71 and chief medical resident at the Nashville Veterans Administration Hospital in 1971-72.
213 The 11th edition of the *Cecil-Loeb Textbook of Medicine,* published in 1963, the successor of *Osler's Principles and Practice of Medicine,* was a 239 cubic inch book.
214 Robert K. Johnston graduated from Vanderbilt University School of Medicine in 1966.
215 Pinson, Barbara. *The Inimitable T.E.B.* Nashville!. October, 1977. Pinson was the wife of a Vanderbilt medical student.
216 Dorothy Mott Brittingham, interview by Dr. Eric Dyer, Nashville, TN. January, 2003.
217 T.E.B., Nashville, Tennessee, to Mrs. H.H. Brittingham, Fort Worth, Texas, January 24, 1971.
218 T.E.B., St. Louis, Missouri, to Dorothy Mott Brittingham, Brightwaters, New York, July 5, 1959
219 Dr. Thomas E. Brittingham IV, interview by author, Nashville, Tennessee, August 13, 2017
220 Dorothy Mott Brittingham, interview by author, Nashville, Tennessee, February 20, 2017
221 T.E.B., Nashville, Tennessee, to Mrs. H.H. Brittingham, Fort Worth, Texas, May 8, 1972
222 Dorothy Mott Brittingham, interview by author, Nashville, Tennessee, November 16, 2017
223 T.E.B., Nashville, Tennessee, to Mrs. H. H. Brittingham, Fort Worth, Texas, July 10, 1966
224 T.E.B., Nashville, Tennessee, to Margaret Brittingham, February 25, 1975.
225 T.E.B., Nashville, Tennessee, to Susan Brittingham, February 26, 1974
226 Dorothy Mott Britingham, interview by Dr. Eric Dyer, Nashville, Tennessee, January, 2003
227 T.E.B., Nashville, Tennessee, to Mrs. H. H. Brlittingham, Fort Worth, Texas August 15, 1966
228 Ibid., May 17, 1967
229 Ibid., February 26, 1968
230 T.E.B. to Margaret Brittingham, October 21, 1973.
231 T.E.B., Nashville, Tennessee, to Susan Brittingham, May 1, 1974
232 T.E.B. to Mrs. H.H. Brittingham, January 8, 1961
233 Ibid., February 9, 1973
234 T.E.B., Nashville, Tennessee, to Don, December 11, 1979
235 Grieve, Robert. British Medical Journal 1953, 671. Hutcheson, 1871-1950, was a British physician.

236 T.E.B. to Mrs. H.H. Brittingham, May 17, 1960
237 Dr. Jean Ballinger, interview by author, April 7, 2016. Dr. Ballinger is a practicing surgeon in Nashville.
238 T.E.B., Nashville, Tennessee to Thomas E. Brittingham IV, April 12, 1976.
239 T.E.B. to Margaret Brittingham
240 T.E.B., Fort Worth,Texas, to Sally Brittingham, Nashville, Tennessee, November 1, 1981, and to Thomas E. Brittingham IV, January 24, 1982.
241 T.E.B. to Sally Brittingham, May 8, 1982
242 Eric Dyer, "Brittingham—A Life in Medicine." (Unpublished manuscript), 2005
243 T.E.B. to Susan Brittingham, October 19, 1971
244 T.E.B., Nashville, Tennessee, letter to "Sam," Administrator, Rutherford County Hospital, Murfreesboro, Tennessee, August 23, 1979
245 Des Prez, R. *Remarks during a 1986 reunion of former house staff of the Department of Medicine*. Reprinted in *Vanderbilt Medicine*, Winter, 1987. Dr. Des Prez knew T.E.B. since they were colleagues on the house staff of New York Hospital, 1950-1954. Des Prez was Chief of Medicine at the Nashville Veterans Hospital during T.E.B.'s years at Vanderbilt.
246 T.E.B. to Dr. Merrill Hicks, Nashville, Tennessee, 1976 or 1977
247 Dr. Rick Davidson, Gainesville, Florida, telephone interview by author, November 1, 2016. Dr. Davidson was a 1972 graduate of Vanderbilt Medical School and interned at Vanderbilt, where he was a colleague of T.E.B. in the Nashville General Hospital Clinic in 1973. He later studied clinical epidemiology at the University of North Carolina and has been on the faculty of the University of Florida Medical School since 1984
248 Dr. Agnes Fogo, interview by author, Nashville, Tennessee, December 30, 2015. Dr. Fogo graduated from Vanderbilt Medical School and is now a Professor of Pathology at Vanderbilt.
249 T.E.B. to Susan Brittingham, October 24, 1971
250 Dr. Stan Graber, interview by author, 10/28/2015, Nashville, Tennessee. Dr. Graber, a cousin of the author, graduated from Vanderbilt Medical School and is now a retired Vanderbilt hematologist
251 DeVita, Vincent et al. *Combination chemotherapy in the treatment of advanced Hodgkin's Disease.* Annals of Internal Medicine 73:881-895. 1970
252 T.E.B. to Susan Brittingham, November 16, 1969
253 Ibid, September 28, 1971
254 Dr. Sanford Krantz, interview by author, Nashville, Tennessee, July 11, 2016. Dr. Krantz was chief of hematology at Vanderbilt before he retired
255 T.E.B. to Susan Brittingham,, May 15, 1972
256 Ibid, July 8, 1972
257 Dr. Alan Cohen, interview by author, March 23, 2016, Nashville, Tennessee. Dr. Cohen trained on the Vanderbilt house staff and is now a retired oncologist

258 Dr. William S. Stoney, interview by author, Nashville, Tennessee, May 1, 2017. Before his retirement, Dr. Stoney was chief of cardiac surgery at St. Thomas Hospital, Nashville, TN.
259 Stoney, W. H. *Pioneers of Cardiac Surgery*. Vanderbilt University Press. Nashville. 2008, 42-43
260 Preston, T.A. *Coronary Artery Surgery. A Critical Review*. Raven Press. New York.1977
261 CASS Principal Investigators and Their Associates. *Coronary Artery Surgery Study (CASS): A Randomized Trial of Coronary Bypass Surgery. Survival Data*. Circulation. 1983;68: 939-950
262 T.E.B. to Susan Brittingham, February 17, 1971
263 Ibid., October 5, 1972
264 T.E.B., Nashville, Tennessee, to Mrs. H.H. Brittingham, Fort Worth, Texas, August 23, 1964
265 T.E.B. to Susan Brittingham, May 1, 1974
266 Dr. Eric Dyer. "Brittingham—A Life in Medicine." (Unpublished manuscriptipt), 2005
267 T.E.B. to Dr. Robert Dunkerly, June 29, 1979
268 John E. Chapman, interview by Dr. Eric Dyer, March 19, 2003
269 Ludmerer, 231
270 Dorothy Mott Brittingham, interview by Dr. Eric Dyer, Nashville, Tennessee, February 2, 2003.
271 Ibid.
272 T.E.B. to Dr. Grant Liddle and Dr. John Oates, September 22, 1979
273 David M. Gaba and Steven K. Howard, *Fatigue among clinicians and the safety of patients*. New England Journal of Medicine 2003; 347:1249-1255.
274 Ludmerer, 297
275 T.E.B. to Dr. Grant W. Liddle, Chairman of Department of Medicine, Nashville, Tennessee, May 22, 1977
276 T.E.B. to William Waller, Vand University Hospital Board of Trust member, Nashville, Tennessee, June 14, 1976
277 Dorothy Mott Brittingham, interview by Dr. Eric Dyer, Nashville, Tennessee, March 5, 2003.
278 Dr. Robert Collins, Department of Pathology, Vanderbilt University School of Medicine, to T.E.B., Nashville, Tennessee, December 18, 1979
279 Dr. Robb Rutledge, interview by Dr. Eric Dyer, Fort Worth, Texas, December 1, 2002.
280 Dr. John Nickell, interview by Dr. Eric Dyer, Fort Worth, Texas, March 1, 2003.
281 Dr. Ed Nelson, Forth Worth, Texas, telephone interview by author, January 4, 2017.
282 Dr. Tom Q. Davis, Colorado Springs, Colorado, telephone interview by author,July 17, 2016.

283 T.E.B., Fort Worth Texas, to Sally Brittingham, Nashville, Tennessee, February 27, 1980
284 Dr. Ed Nelson, January 4, 2017.
285 T.E.B. to Sally Brittingham, February 17, 1980
286 Dr. Tom Q. Davis, July 17, 2016.
287 T.E.B. to Sally Brittingham, May 3, 1980
288 T.E.B. to Sally Brittingham, May 18, 1980
289 T.E.B., Fort Worth,Texas, to Dr. Dykes Cordell
290 T.E.B. to Sally Brittingham, February 2, 1981
291 T.E.B., Fort Worth, Texas, to Thomas E. Brittingham IV, Knoxville, Tennessee, February 23, 1982.
292 Ibid., April 24, 1982.
293 Dr. David Rogers, Princeton, New Jersey to T.E.B., Fort Worth, Texas, February 23, 1981
294 T.E.B. to Thomas E. Brittingham IV, June 26, 1982 and August 6, 1982
295 T.E.B. to Sally Brittingham, June 11, 1981
296 T.E.B. to Thomas E. Brittingham IV, July 24, 1981
297 T.E.B., Fort Worth,Texas, to Dr. Jeanne Ballinger, Nashville, Tennessee, January 8, 1983
298 T.E.B., Fort Worth, Texas, to Dr. James L. Fletcher, Augusta Georgia, March 19, 1984
299 Dorothy Mottt Brittingham, interview by author, Nashville, Tennessee, November 16, 2017
300 T.E.B., to Dr. George McIlheran, Chairman of the Medical Staff Quality Assurance Committee at Harris Hospital, July 17, 1984.
301 Dr. Robb Rutledge, interview by Dr. Eric Dyer, Fort Worth, Texas, January 15, 2003.
302 Dr. John S. Alexander, interview by Dr. Eric Dyer, Fort Worth, Texas, February 22, 2003.
303 Bryan, C. S., Persistent fever in three otherwise healthy patients, *Consultant*, February, 1986, 167-185.
304 Telephone interview by author with Dr. Tom Q. Davis, Colorado Springs, Colorado, July 17, 2016.
305 Dr. Roger Des Prez. Vanderbilt Medicine, Winter 1987: *A Tribute to Dr. Thomas E.Brittingham*, during a 1986 reunion of former house staff of the Vanderbilt Department of Medicine.
306 Dr. Robb Rutledge, December, 2002
307 Dr. John Sergent, interview by author, Nashville, Tennessee, October 14, 2015
308 Dr. Karl Vandevender, interview by author, Nashville, Tennessee, January 29, 2016
309 Dr. Jimmy Sullivan, interview by author, Nashville, Tennessee, October 19, 2015.

310 Quoted by Bryan, Charles S. The art of Medicine—Osler redux: the American College of Physicians at 100. www.thelancet.com. 2015;385:1720-1721.
311 Dr. David Robertson, interview by author, Nashville, Tennessee, October 18, 2015.
312 Dr. John (Dick) Dixon, interview by author, Nashville, Tennessee, April 1, 2016.
313 Ibid.
314 Dr. Clif Cleaveland, telephone interview by author. Chattanooga, TN, November 21, 2015.
315 Dr. David Dodson, interviews by author, December 8, 2015, Manchester, Tennessee, and January 15, 2016, Chattanooga, Tennessee.
316 Gunderman, R. Hospitalists and the Decline of Comprehensive Care. *N. Engl. J. Med* 2016: 1022-1013.
317 Ludmerer, K.M., p.131
318 Dr. William Schaffner, interview by author, Nashville, Tennessee, February 16, 2016.
319 Smith, L. G. Medical professionalism and the generation gap. *Am. J. Medicine* 2005:11, 439
320 Bryan, C.S. *Some of my Teachers*. Farewell address (and introductory lecture on Professionalism) to first year students at the University of South Carolina School of Medicine. 2008. Dr. Bryan served as an intern and resident at Vanderbilt during T.E.B.'s tenure, then as an infectious disease fellow at the Nashville VA Hospital under Dr. Roger Des Prez. He became Chief of Medicine at University of South Carolina in 1992, where he also was Director of the Center for Bioethics and Medical Humanities
321 Smith, L.G., 440
322 Zenke, R. et al. *Generations at Work*. American Management Association. New York. 2000
323 Ludmerer, K.M. *Let Me Heal,* 266
324 Ibid., 267
325 bid., 268
326 Ibid, p. xiii
327 Dr. John Sergent, interview by author, Nashville, Tennessee, October 14, 2015
328 Dr. James R. (Bo) Sheller, interview by author, Nashville, Tennessee, October 15, 2015
329 Rosenthal, D.I, Verghese, A. Meaning and the Nature of Physicians' Work. *N. Engl. J. Med.* 2016:375;1813-1815.
330 Dr. Henry Jennings, interview by author, Nashville, Tennessee, April 29, 2017
331 Dr. Seth Cooper, interview by author, Nashville, Tennessee, February 24, 2016. Dr.
332 Dr. Frank Boehm, interview by author, Nashville, Tennessee, January 27, 2016
333 Dr. David Orth, interview by author, Nashville, Tennessee, April 22, 2017. Dr. Orth served his internship and residency on the Osler Service at Johns Hopkins

and subsequently was Chief of Endocrinology at Vanderbilt for several years. Dr. Liddle was a mentor of both Dr. Orth and the author

[334] Oshinsky, D. *Bellevue—Three Centuries of Medicine and Mayhem at America's Most Storied Hospital*. Doubleday, a division of Penguin Random House LLC. New York. 2016

[335] Rosenthal, Elizabeth. *An American Sickness—How healthcare became big business and how you can take it back*. Penguin Press. New York. 2017

[336] White, W. Aequanimitas: Osler's Inspirational Essays. *Bull. Hist. Med.* 6:820-833, 1938. Quoted by Bryan, C.S. Aequanimitas Revisited. *J. S. Cal. Med Assn* 1995, August, 357

[337] Dr. John Leonard, interview by author, Nashville, Tennessee, February 25, 2016

[338] Dr. Clif Cleveland,, May 13, 2015

[339] Dr. Alexander McLeod, telephone interview by author, Nashville, Tennessee, September 29, 2015

[340] John E. Chapman, interview by Dr. Eric Dyer, Nashville, Tennessee, March 19, 2003.

www.ingramcontent.com/pod-product-compliance
Lightning Source LLC
Chambersburg PA
CBHW031944170526
45157CB00002B/387